Rural Retirement Migration

THE SPRINGER SERIES ON
DEMOGRAPHIC METHODS AND POPULATION ANALYSIS

Series Editor
KENNETH C. LAND
Duke University

In recent decades, there has been a rapid development of demographic models and methods and an explosive growth in the range of applications of population analysis. This series seeks to provide a publication outlet both for high-quality textual and expository books on modern techniques of demographic analysis and for works that present exemplary applications of such techniques to various aspects of population analysis.

Topics appropriate for the series include:

- General demographic methods
- Techniques of standardization
- Life table models and methods
- Multistate and multiregional life tables, analyses and projections
- Demographic aspects of biostatistics and epidemiology
- Stable population theory and its extensions
- Methods of indirect estimation
- Stochastic population models
- Event history analysis, duration analysis, and hazard regression models
- Demographic projection methods and population forecasts
- Techniques of applied demographic analysis, regional and local population estimates and projections
- Methods of estimation and projection for business and health care applications
- Methods and estimates for unique populations such as schools and students

Volumes in the series are of interest to researchers, professionals, and students in demography, sociology, economics, statistics, geography and regional science, public health and health care management, epidemiology, biostatistics, actuarial science, business, and related fields.

The titles published in this series are listed at the end of this volume.

Rural Retirement Migration

by

David L. Brown and Nina Glasgow

Department of Development Sociology
Cornell University
Ithaca, New York, USA

With Contributions from:

Laszlo J. Kulcsar
Benjamin C. Bolender
Marie-Joy Arguillas

 Springer

David L. Brown
Department of
Development Sociology
Cornell University
Ithaca, New York
USA

Nina Glasgow
Department of
Development Sociology
Cornell University
Ithaca, New York
USA

ISBN: 978-1-4020-6894-2   e-ISBN: 978-1-4020-6895-9

Library of Congress Control Number: 200892199

Printed on acid-free paper

9 8 7 6 5 4 3 2 1

springer.com

For our parents, Ethel and Willard Glasgow and Bernice and Lawrence L. Brown Jr., who inspired our interests in aging, and for Glenn V. Fuguitt and Calvin L. Beale, two productively aging rural demographers who have taught us much

CONTENTS

ACKNOWLEDGMENTS

This book began to take shape in 2001 when we embarked on a research project that combined our long-time separate interests in social gerontology and migration into an integrated study of older persons who move to rural retirement destinations and the communities to which they move. Large, multi-level research projects such as this are not easily funded, and we were fortunate to obtain support from a number of organizations and agencies. The survey research part of the study was supported by a competitive grant from the U.S. Department of Agriculture's National Research Initiative. The county-level analysis of rural retirement destinations was supported through our participation in USDA multi-state research project W-1001, "Rural Population Change," and the community case studies were funded by Cornell University's Polson Institute for Global Development. We thank all of these organizations, but especially our W-1001 colleagues for their intellectual support and encouragement. Writing began in earnest in 2006 during our sabbatical leave at the University of Newcastle upon Tyne's Centre for Rural Economy. We thank Professor Neil Ward and our other colleagues at CRE for their warm hospitality and intellectual stimulation.

Many individuals played a part in conducting this research and producing the present volume. Laszlo J. Kulcsar and Benjamin C. Bolender co-authored Chapter 3, and Marie-Joy Arguillas co-authored Chapter 5. Sarah Giroux copy-edited the entire manuscript and made many helpful suggestions for improving the writing. Yasamin Miller and Eric Nesbit of the Cornell University Survey Research Institute helped us design the panel survey of older persons living in rural retirement destinations. Max Pfeffer, John Cromartie, Ken Johnson, Glenn Fuguitt, and David McGranahan provided county-level data that supported our analyses of rural retirement destination communities. Ania Choike and Benjamin Allen, our undergraduate assistants at Cornell, prepared community profiles that were helpful in preparing to conduct the case studies. We wish to thank the nearly 800 older persons who responded to our survey. Without their participation our study would have come to nothing. Thanks also go to the community leaders in the four case study communities who were so open and candid during our interviews. Scott Sanders prepared the figures and maps, and Linda Warner typed the tables. Deborah Patton indexed the book. Evelien Bakker and Bernadette Deelen, our editors at Springer, provided helpful advice and continuing support.

LIST OF FIGURES

CHAPTER 1

RURAL RETIREMENT MIGRATION: PAST, PRESENT AND FUTURE

INTRODUCTION

Why Write a Book on Rural Retirement Migration?

This book examines the migration of older persons to rural retirement destination communities in the United States. It focuses both on the older in-migrants themselves and the communities in which they settle. The purpose of the research is to gain a fuller understanding of the process of rural retirement migration at both micro- and macro-levels of social organization. The study is motivated by the following questions:

- What is a rural retirement destination community? Why are some rural areas more likely to become a destination for older in-migrants and to maintain this status over time?
- Who moves to rural America during older ages and how do these persons compare with similar-aged longer-term residents? What social and economic resources do in-migrants contribute to rural retirement destinations?
- What social processes motivate older persons to move to rural communities and what steers them to particular locations?
- How do in-migrants become socially integrated in their new communities? What personal attributes increase or decrease the likelihood that older in-migrants will become socially integrated in their new communities?
- What forms of social integration are most likely to contribute to in-migrants' health and well-being as they age-in-place in rural retirement destinations?
- What is the nature of in-migrants' social and institutional participation in their new communities?

- From a community perspective, what are the pros and cons of rural retirement migration? What are the short- and longer-term implications of being a destination for older in-migrants?

Background and Importance of Population Aging

Population aging is one of the world's most dramatic and widespread demographic phenomena. Over 400 million people are now age 65 and older, about 7 percent of all persons on the earth. This is an increase from 5 percent in 1950, or almost 300 million older persons. While the percent 65 and above varies greatly across the world's regions, it is increasing everywhere, even in sub-Saharan Africa. This being acknowledged, the magnitude of population aging is significantly greater in the U.S. and more developed parts of the world (United Nations Population Division, 2003). For example, in Europe and the U.S. fully 15 and 12 percent, respectively, of their populations were aged 65 or older in 2000 compared with slightly over 4 percent across most of Asia (Federal Interagency Forum on Aging Related Statistics, 2004). Over the last 50 years, America's older population grew from 12 million to 35 million, with the highest rate of increase coming at age 85 and above.

The social and economic implications of population aging are far reaching. As Fuguitt, Brown and Beale (1989, p. 105) observed,

> The age–sex composition of a population shapes community needs and demands for goods, services and economic opportunities, as well as patterns of consumption, life style and social relationships. It is through changes in age–sex composition that demand shifts associated with population growth and decline are most closely articulated. The age–sex composition of a population imposes requirements and limitations on each of its institutions.

Health care, transportation, housing, income security, cultural facilities and social and civic engagement are directly affected by age structure and changes therein. Human communities are institutionalized solutions to solving the challenges of everyday life, and where one lives matters! Since most older persons are permanent or semi-permanent residents of particular geographic locales, satisfying their everyday needs is contingent on the communities in which they live. This study focuses on rural retirement destinations, a particular social environment where older Americans are concentrating and where future cohorts of older persons will almost certainly continue to live in disproportionate numbers.

As indicated above, approximately 12 percent of Americans are currently aged 65 or older, with this share increasing to at least one fifth when the baby boom enters retirement age (Federal Interagency Forum on Aging

Related Statistics, 2004). While older persons are less likely to migrate than their younger counterparts (U.S. Census Bureau, 2003), a disproportionate share of those who do move are likely to head for rural destinations. Thus, aging patterns in rural areas will be of increasing interest in the near future (Fuguitt, Beale and Tordella, 2002).

The book examines both the micro- and macro-aspects of rural retirement migration. It focuses on older in-migrants themselves at the same time as it examines the destination communities in which they settle. It provides an empirically based assessment of factors associated with the development of rural retirement destinations and the community-level impacts associated with attracting older retirees. At the micro-level, we examine the social and economic dynamics of older in-migration to rural areas and the process by which older in-migrants establish social relationships, organizational memberships and community involvements in their destination communities. Additionally, we evaluate the outcomes for migrants that are associated with various forms of social integration. At the macro-level, we examine the socio-demographic and institutional impacts associated with being a destination for large numbers of older in-migrants. While written from a sociological and social demographic perspective, the book has broad multi-disciplinary appeal in sociology, social gerontology, geography, planning, community development, social work and public health. We provide an empirical framework for considering the determinants and consequences of retirement migration growth in rural environments, and we enhance understanding of why some in-migrants are more successful than others in becoming socially integrated, as well as the implications of such social integration or lack thereof.

POPULATION AGING IN THE U.S. IN THE 21ST CENTURY

The Dynamics of Population Aging

As indicated above, population aging is a world-wide phenomenon. Populations age partly as a result of declining mortality, but the effect of declining mortality on a population's age structure is contingent on a nation's level of development. In less developed nations, where fertility remains relatively high, mortality declines are most likely to occur at the younger ages. In that context, declining mortality tends to result in a younger population. However, even though fertility remains relatively high, fertility declines in developing nations are increasing the percentage of the population in the older age groups. Even though only 3 percent of the population in Sub-Saharan Africa is 65 and older at the present time, for example, this percentage has begun to rise in response to declining fertility. The United Nations predicts that over 5 percent of Sub-Saharan Africa's population will be 65 or older by mid-century compared with less than 3 percent in 2000 (United Nations Population Division, 2003).

In a low mortality rate society like the U.S., on the other hand, population aging is associated with declining mortality because the majority of all deaths occur at ages 65 and older. As Weeks (2005, p. 342) points out, "At a life expectancy of 65 with replacement-level fertility, the average age is 38; the average age rises to 41 as life expectancy increases to 80. The percentage of the population aged 65 and older increases from 15 percent to 24 percent...."

Age-selective migration also affects age structure. Age-selective population movement adds or subtracts persons from specific age groups, directly affecting age composition. But migration's impact can also be indirect and experienced over a longer time horizon. For example, chronic net out-migration of persons in their child-bearing years not only subtracts such persons from the younger part of a population's age structure, but diminishes the population's future reproductive capacity resulting in further aging.

Rural Population Aging in the U.S.

The percent of the *rural* population at ages 65 and older (15 percent) exceeds that of the country as a whole (12 percent) (Economic Research Service-USDA, 2007). About seven and a half million rural persons were aged 65 and older in 2004, an increase of 2.3 percent since 2000. As Fuguitt and Beale (1993) demonstrated, changes in the absolute and relative size of the rural population can be decomposed into the underlying components of natural increase and net migration. In this case, because the focus is on growth of the older population, natural increase does not involve actual births. Rather, the survival of persons aged 55–64 into the 65 and over category is a proxy for "births", which when combined with deaths over the decade provides a measure of natural increase. Migration, the other component of change, has both direct and indirect and short- and longer-term impacts on rural aging. Older persons move into or out of areas, which directly affects age structure. In contrast, chronic out-migration of younger persons ages the population both directly and over the longer-term through its negative effect on fertility. Research during 2000–2004 by the Economic Research Service-USDA (2007) used the Fuguitt–Beale components of change framework to show that natural increase hit a low point during this time, reflecting the small Depression-era birth cohorts entering old age in the first half of the decade. They reported, however, that the propensity to move to rural areas continued to be high among empty nesters and retirees. With respect to the present study, we see that retirement migration is contributing to rural aging and has been doing so for at least 30 years (Fuguitt and Beale, 1993;

Johnson, Voss, Hammer, Fuguitt and McNiven, 2005; Lichter, Fuguitt, Heaton and Clifford, 1981). The ERS researchers also noted that rural to urban migration among young adults outnumbered the counter-stream, hence also contributing to an older rural population. They pointed out that even during periods of above average rural growth, more young adults leave rural areas than are gained through in-migration.

Rural Aging and the Aging of the Baby Boom

During the height of the baby boom, between 1956 and 1964, over 4 million children were added to the U.S. population annually. This level of fertility was unprecedented in American demographic history, but it did not signal a return to extremely large families characteristic of the 19th century. Rather, the average number of children born to couples during this time increased from a little more than two to about three and a half as women married and had their first child earlier, as more children were born to older women and as the rate of childlessness declined. As a result of both higher fertility during the 1950s and 1960s and increased longevity, an unprecedented number of baby boom survivors are now on the cusp of entering older age.

The aging of the baby boom will differentially affect population change across the nation's regions and in both urban and rural areas. As Plane (1992) and Plane and Rogerson (1991) have demonstrated, the movement of baby boom cohorts through the age structure partially explains population redistribution from the Northeast to the South and West during the 1980s. In addition, Nelson, Nicholson and Stege (2004, p. 540) have shown that "fluctuations in nonmetropolitan population change are associated with the sign and magnitude of cohort shifts for the peak baby boom years." Their analysis showed that region-wide trends can mask internal variation in population growth and decline. While the Northeast and Midwest declined overall in population during the late 1970s and 1980s, for example, nonmetropolitan parts of these regions experienced growth because baby boomers moved to these areas in much larger than expected numbers. When boomers reach retirement age they will become less sensitive to local labor market conditions, and their preferences for amenities, leisure activities and proximity to children and grand children will undoubtedly increase (Longino, 1992). The geographic mobility of baby boomers is likely to have a growing impact on population redistribution in the U.S. both within and between regions and between rural and urban areas. One should be cautious, however, about predicting the baby boom's geographic behavior. Like any age cohort, the baby boom is diverse with respect to life experiences and life values (American Association of Retired Persons, 1999). Moreover, as Haas and Serow (2002) have indicated, the baby boom's geographic mobility will be contingent on the nation's macro-economic situation at the time they retire and on the institutional environment, including the Social Security system's solvency.

As the baby boom reaches older age, a large number of retirees are likely to consider migrating and many who do move are likely to follow residential preferences for rural destinations formed earlier in their lives (Brown, Fuguitt, Heaton and Waseem, 1997). While the size of this older migration stream is uncertain at this time, it may be substantial, at least in comparison with the populations currently residing in rural destination communities. It is also likely that a substantial share of retired boomers who choose to move to rural areas will head for those areas that have already become established as centers for retirement living. These rural retirement destination communities are the focus of this book. Our examination of rural retiree in-migrants and the communities to which they move provides important information for considering the future impacts of the aging of the baby boom in rural America during the coming decades.

GEOGRAPHIC MOBILITY AMONG OLDER PERSONS IN THE U.S.

Geographic mobility of older persons became a focus of scholarly attention around 1980 when Longino and Jackson edited a special issue of *Research on Aging* that concentrated on this topic. Writing in that issue, Everett Lee (1980, p. 135) commented that "... migration of the elderly has increased with time.... What we have so far witnessed is only the beginning of the movement." Census data, however, indicate that the rate of residential mobility among older persons is relatively low compared with younger persons (about 23 percent of persons aged 65 and older changed residence between 1995 and 2000). Younger persons were more than twice as likely to move as persons aged 65 and older. Moreover, most older persons who change their residence move locally, with only about 9 percent of older persons moving across county lines between 1995 and 2000 and a mere 4 percent migrating across state boundaries (U.S. Census Bureau, 2003).

Longer distance migration among older persons reflects more general patterns of regional population redistribution in the U.S. (U.S. Census Bureau, 2003). Similar to the nation's overall pattern of internal migration, older migrants tend to move from the Northeast and Midwest to the South and West. The South, and especially the South Atlantic sub-region, experienced the largest net gain of older persons of any region through migration during the 1990s, though the Mountain West was also a big gainer. In contrast, the Pacific sub-region switched from net in-migration to net out-migration during the decade. California alone lost 34,000 older persons through migration. Many of these persons moved to Arizona, Nevada and other parts of the interior West. The Middle Atlantic sub-region experienced the largest net loss of older persons through out-migration during the 1990s. New York experienced the greatest loss—114,000 older persons during this time—with Illinois and Michigan also experiencing substantial net out-migration.

Migration and the Life Course

Residential mobility varies within the 65 and older category itself. Ironically, the oldest-old are more likely to move than their younger-old counterparts. Mobility at the oldest ages is often associated with chronic illness, the onset of disability and/or the loss of a spouse or some other adverse life course event. A large portion of these late life moves transfer people from independent living to assisted living facilities, continuing care retirement communities or nursing homes. While many moves among the oldest-old are local, others cross state and county lines as older persons relocate to be closer to their adult children and/or the more extensive health care services offered in large urban areas.

As illustrated by the above trends, migration and other forms of residential mobility often accompany life course transitions, including transitions that occur later in life. Eugene Litwak and Charles Longino (1987) have proposed a "developmental" model of older migration that recognizes the association between life course events and transitions and changes in residential location. They identify three kinds of moves among older persons: (a) amenity-driven residential relocations that occur close to the time of retirement; (b) changes of residence that occur with the onset of moderate forms of disability, especially in combination with the loss of one's spouse; and (c) moves associated with major forms of chronic disability. With this proposed as a developmental framework, the authors see the three types of moves as taking place in successive stages as older persons progress through the later stages of the life course. In the first stage, retirees and/or pre-retirees make leisure-oriented moves to amenity-rich communities. The second stage often returns older migrants to their previous communities or to places where their children live. The final stage occurs when older persons move from independent living to an institutional setting. This life course framework has important implications for the study of migration among the rural aging. In particular, it raises questions about the length of time older in-movers can be expected to spend in rural retirement destinations, their health and well-being during this time and their needs and contributions while residing in rural America. It also begs the question of how closely the proposed developmental stages of older migration conform to empirical reality.

Five Macro-Societal Factors that Make Migration of Older
Persons More Likely

The rate of interstate migration among older persons has remained very stable at about 4.5 percent over the past three decades (Longino, 2006). In contrast, the rate of interstate mobility among younger persons has declined from 9.9 percent during the 1970s to 8.4 percent between 1995 and 2000.[1] The fact that migration among older persons has remained steady at the same time as it has declined among younger persons testifies to the trend's strength and stability. Migration is less prevalent among older persons than their younger counterparts because older persons have fewer years over which to realize the returns from moving. However, at least five major social forces make geographic mobility a reasonable proposition for older persons who are interested in moving.

To begin with, life expectancy at age 65 has increased dramatically during the last 30 years. In 2001, people aged 65 could expect to live 18 more years, which is 3 years longer than similar aged persons could expect to live in 1970 (Federal Interagency Forum on Aging Related Statistics, 2004).[2] Life expectancy is a measure of the overall health of the population and increased longevity not only indicates more years, but more healthy years. Increased life expectancy means that older migrants can expect to enjoy positive returns from migration over a significant time horizon.

Related to the above discussion of increased longevity, health among older persons has increased during recent years and chronic disability has declined. Data from the National Center for Health Statistics show that age-adjusted death rates for all causes of death among people aged 65 and older declined by 12 percent between 1981 and 2000. Chronic disability declined from 25 percent to 20 percent among people aged 65 and older during this time (Federal Interagency Forum on Aging Related Statistics, 2004). Moreover, while rural areas still have higher age-specific mortality rates from most causes (Morton, 2004), these improvements over time seem to characterize both urban and rural areas. For example, research by Johnson (2004) showed little rural disadvantage in disability among adults. Improved health and reduced disability mean that older persons are better able to participate in recreational, organizational and civic activities. Consistent with our observations about longevity, healthy older persons are better able to justify moving because they can expect to be actively involved in their new communities for a significant period of time after their arrival.

One result of enhanced health and lower disability among older persons is that an increasing share of older persons continue to live independently for a longer time than was true in the past. In 1985, 54 people per 1000 aged 65 and older resided in nursing homes. This rate declined to 43 per 1000 in 1999. Even at the oldest ages, the rate of nursing home residence has declined from 220 per 1000 in 1985 to 183 per 1000 in 1999 (U.S. National Center for Health Statistics, 2000).

Research shows that labor force participation declined from 76 percent to 50 percent among men aged 62–64 between 1963 and 2003. During the same period, women's labor force participation at these ages increased from 29 to 39 percent. Even though women's labor force attachment has increased, less than half of older persons are closely tied to the local economy by the time they reach their early to mid-60s. As a consequence of looser ties to the local labor market, older persons are more able to consider changing their residential location. The decline in labor force participation by men is associated with employer-provided pensions, the reduced age of eligibility for Social Security and greater wealth accumulation. Data from the Current Population Survey (CPS), for example, show that the wealthiest two quintiles of America's older persons—those most likely to migrate at age 60 and above—derive over one third of their income from assets and pensions (U.S. Census Bureau, 2003b).[3]

Finally, numerous data sources show that economic security has risen among America's older population thus increasing their choices and options, including their residential choices and options during retirement. CPS data show that poverty among older persons has declined from 15 percent in 1974 to about 10 percent in 2002.[4] During the same time, median household income among the older population increased from $16,882 to $23,152.[5]

Rural Retirement Migration

Social scientists have recognized that particular rural regions became net recipients of older in-migration as early as the 1970s (Sofranko, Fliegel and Glasgow, 1983). The phenomenon has continued since then and rural retirement migration has become a research focus in sociology, geography, demography, social gerontology and other social science disciplines. Wake Forest University's Reynolda Gerontology Program maintains a bibliography that lists almost 500 research citations on retirement migration, many of which focus on migration to rural destination communities (Wake Forest University, 2007).

Rural retirement migration has been of particular interest because regardless of the overall direction of urban to rural migration—positive in the 1970s, the early 1990s and the 2000s and negative in the 1980s—more older persons have moved to rural areas than in the opposite direction in each of those decades (Fulton, Fuguitt and Gibson, 1997). Net in-migration of older persons played an important role in the rural population turnaround of the 1970s (Glasgow, 1980) and, since then, has continued to be associated with relative prosperity and community vitality within rural America. Counties with higher than average net in-migration at older ages are among the most rapidly and consistently growing types of rural areas. Retirement destination counties, by definition, attract older migrants, but they also attract working-age persons who obtain service, retail and construction jobs induced by the in-flow of retirees (Johnson and Fuguitt, 2000). As will be discussed in Chapter 2, retirement migration is seen as an engine of economic development and many states of the U.S. have designed programs with the explicit goal of recruiting retirees (Reeder, 1998).

Migration, including older migration, is typically selective of persons with particular characteristics. Past research has consistently shown status differences between older in-migrants and longer-term residents, with in-migrants being positively selected by comparison (Biggar, 1980; Glasgow, 1995). As Longino and Bradley (2006) pointed out, moving is costly and thus screens out persons who cannot afford to move. They also observed that retirees move soon after retiring while their assets are still largely intact. Retirement income is highly portable, and older migrants can take their assets, pensions and Social Security with them regardless of where they choose to live. This makes retirees attractive to areas seeking an infusion of outside capital—hence the recruitment schemes mentioned above.

Rural communities also incur costs when older in-migrants arrive. For example, concern has been voiced that retirement migration will elevate the use of public services, thus experiencing heightened fiscal stress. Proponents of this view argue that because many older in-movers originate in urban areas with more extensive service providing environments, they will have a stronger inclination to use services than comparable persons who have lived in rural areas for long periods of time (Anderson and Newman, 1973). Glasgow (1995), however, showed that migration per se does not predict elevated use of most public services. More generally, research shows that even though some costs are incurred, rural places benefit economically from retirement in-migration (Glasgow and Reeder, 1990; Stallman, Deller and Shields, 1999). But as will be discussed in Chapter 6, very little research has examined the social and institutional effects of older in-migration.

International Experiences

Retirement migration is not strictly, or even primarily, an American phenomenon. There are well-established migration streams at older ages between northern Europe and Spain, Italy, Cyprus and a variety of other southern European destinations.[6] Warnes (2001), for example, has conducted extensive research on the international migration of British retirees. He explains that the increased international dispersal of British expatriate retirees in southern Europe is a result of globalization, improved transportation infrastructure and greater international experience during the working lives of persons who are now retired (King, Warnes and Williams, 2000). Similar to the American situation, amenity-migrants often indicate that they have vacationed previously in the destination region (Rodriguez, Fernandez-Mayorales and Rojo, 1998). But, as King and Patterson (1998) observed, migration processes and life histories of British retirees living in Tuscany are extremely heterogeneous. In fact, many English retirees in Tuscany did not move there directly from England, nor do they consider themselves to be retired. Moreover, similar to the U.S., older people in other developed countries participate in seasonal migration. For example, much research has focused on the cyclical movement of older Canadians to Florida. This research pays particular attention to the effects of national differences in health care systems on the temporal nature of this migration stream (Marshall, Longino, Tucker and Mullins, 1989).

Not all European retirement migration is from north to south, nor even to Spain, Italy or other relatively high-income countries. Illes (2005) reported that older immigration to Hungary has been growing since the mid-1990s. He concludes that non-Hungarian amenity seekers draw sufficient pensions from their countries of origin for a high standard of living. In contrast, pensioners who arrive in Hungary to reunite with their families appear to have far less economic security and are likely to become dependent, to some extent, on the Hungarian state.

Similar to the U.S. situation, some European retirement migration involves internal redistribution rather than international movement. Champion and Sheppard (2006) demonstrated that net in-migration to rural England at ages 60 and older is strongly positive, as it is at pre-retirement ages. Accordingly, rural England can expect population aging over the next decades both as an effect of older in-migration and because of aging-in-place among retirees and pre-retirees. Atterton (2006) has demonstrated that population aging is especially marked in Britain's coastal regions. She points out that population aging has traditionally been considered to be a problem in Britain, "a pensions and care issue," but she indicates that older people have much to contribute to their communities. This is consistent with the position taken by Le Mesurier (2006) in Age Concern England's recent publication, *The Ageing Rural Countryside*. He concludes that social participation by older people in rural communities belies stereotypes of dependency and age. Productive rural aging has even been recog-nized by the *Sunday Times*. In a piece entitled "Bucolic Bliss," Fred Redwood (2006, p. 6) concluded, "Once dismissed as dull, country towns are buzzing with affluent oldies." Rural retirement in-migration has also been reported in Australia, especially in "life-style regions" (Garlick, Waterman and Soar, 2006). As a result, the Australian Local Government Association has mounted a campaign to promote timely action by local governments in response to population aging.

It appears that few of the issues raised in this book are exclusive to the U.S. Therefore, rural retirement migration in the United States should be considered in an international context. This diverse phenomenon is an element of demographic and socio-economic change in developed nations throughout the world and it is likely to accelerate in the future as the world's baby boomers enter retirement age.

THE CORNELL RETIREMENT MIGRATION PROJECT

As indicated above, most previous research on retirement migration has focused on the phenomenon's demographic dynamics and its economic and fiscal effects at the community level. Surprisingly little research has focused on the adjustment of older in-migrants or the community-level social and institutional impacts even though the need for such research has been recognized for quite a while (Rowles and Watkins, 1993). Moreover, most previous research has either focused on the migrants *or* the destination communities. As little or no work has sought to merge these micro- and macro-domains in an integrated program of research, our research seeks to fill this gap. We examine the nature of social relationships older in-migrants form among themselves and with longer-term residents, and investigate the association between in-migrants' social integration and well-being, especially their health. In addition, we illuminate community-level effects associated with receiving relatively large numbers of older in-migrants. We identify the social and organizational domains most directly affected by older in-migration, the contributions older in-migrants make to these civic and institutional domains and the community-level challenges faced by retirement destinations.

Challenges of Conducting Multi-Method Research

When we began this research in 2001, we realized that conducting multi-method research presented a number of difficult challenges. We needed a clear conceptual framework, an integrated research design and a substantial amount of funding to complete the entire project. Most importantly, we needed a conceptual framework that was broad enough to embrace the migrant integration process while at the same time enabling us to examine the effects of in-migration on social institutions in destination communities. We were guided by Brown's (2002) statement that migration and community are interrelated social processes. The fundamental insight contained in his work, which helps to shape the present research, is that one cannot understand migration without examining the social context in which it is embedded. Similarly, one cannot understand community structure and change in contexts where migration is occurring without examining the demographic, social and economic consequences of migration. In other words, migration affects and is affected by the social relationships that combine to produce community.

We also needed a framework to guide our analysis of migration and social integration. Pillemer, Moen, Wethington and Glasgow's (2000, p. 8) conceptualization of social integration as "the entire set of an individual's connections to others in his or her environment" provided guidance for this part of the analysis. To say that an individual is highly integrated in this sense means that adaptation of older in-migrants is contingent on being embedded in a network of proximate relationships, social affiliations and community organizations (Glasgow and Sofranko, 1980).

Our second challenge was that examining both micro- and macro-level questions in the same study requires a multi-method research strategy. Moreover, since migration and migrants' adaptation are time varying processes, we needed longitudinal data at both the micro- and macro-levels. Our research design includes three linked data sources. We conducted a two-wave panel survey of 368 older in-migrants in 14 rural retirement destinations and a matched sample of 420 longer-term older persons in the same counties to examine the processes of retirement migration and migrant adaptation.[7] The first wave was conducted in 2002 followed by the second wave in 2005. The telephone survey included a full battery of questions on migration, social relationships, social participation, health and disability and demographic and economic characteristics. The data file also included geographic identifiers so that survey data could be linked with county-level secondary data.

While the panel survey contained a wealth of information on the in-migrants themselves, it contained no information about the social, economic and institutional contexts older in-migrants encounter when they move to rural retirement destinations. We used county-level census data and in-depth case studies to place the survey respondents in a social and economic context and to examine how older in-migration affects local institutions in retirement destinations. Case studies were conducted in 4 of the 14 rural retirement counties where the panel survey had been conducted. We conducted one case study in each of the nation's four major geographic regions so that we could examine similarities and differences in the community-level effects of retirement in-migration across distinctly different regional contexts. We interviewed elected and appointed officials, organizational leaders, business owners and aging-service providers to gain an appreciation of the opportunities and challenges rural communities experience when faced with older in-migration. In addition, we held face-to-face interviews with some of the older in-migrants who had participated in both waves of the panel survey. These interviews provided more nuanced information about the nature of their social relationships and factors that constrain and facilitate their community involvement.

The social and economic context of rural retirement migration was established by examining county-level data from the U.S. census and other sources for 1980 through 2000. These analyses described rural retirement destinations and compared them with other kinds of rural environments. Because of the comparative nature of this ecological analysis we were able to consider the kinds of social and economic contexts that attract older in-migrants and the community characteristics that facilitate social participation. In addition, because retirement migration has a strong positive effect on rural population change, we were interested in examining the rural retirement destinations in their own right, not simply as social environments for older in-movers. We wanted to know what these places were like, how they compared with other kinds of rural areas and what county attributes contributed to becoming, and maintaining status as, a rural retirement destination.

The third challenge we faced was obtaining sufficient funding to conduct the entire project. While we articulated the entire program of macro- and micro-level research in our initial proposal, we were unable to obtain sufficient funding to complete all of the work from a single source. In 2001, we received a competitive grant from the U.S. Department of Agriculture's National Research Initiative to examine the social integration and well-being of older in-migrants to rural retirement communities. Research on rural retirement destinations themselves was supported through our participation in USDA multi-state research project W-1001, *Rural Population Change*, and the four in-depth case studies were funded by a small grant from Cornell's Polson Institute for Global Development.

ORGANIZATION OF THE BOOK AND CHAPTER SUMMARIES

The book's analytical core is comprised of five chapters, two that focus on older in-migrants themselves and three that focus on rural retirement communities. In Chapter 2, "Retirement Destinations in the Countryside," the concept of rural retirement destination (RRD) is defined as a nonmetropolitan county with 15 percent or higher in-migration at ages 60 and older. Retiree settlement in these places is into the general community, not into planned retirement communities. Older in-migrants join longer-settled residents, some of whom are also aged 60 and older and who are aging-in-place. Older in-migration contributes to population aging, but because it often induces a general pattern of economic development and in-migration at younger ages, RRDs are not aging as rapidly as many other types of rural communities. Chapter 2 examines the geographic location of RRDs, and we find that, while disproportionately in the South and

West, they are also in the Upper Midwest and in New England. A profile of RRDs in comparison with other types of rural counties is examined and the data show that RRDs are generally larger and more advantaged with respect to population growth and net in-migration, as well as measures of socio-economic status and economic well-being. They are slightly less racially diverse, and, while slightly older than other rural areas, they are considerably younger than farming dependent areas, areas that are experiencing natural population decrease, or those that are losing population.[8]

The chapter then reviews research on the demographic and economic impacts of retirement migration, which reveals a general conclusion that, on balance, communities benefit from attracting older persons. Finally, because research shows that retiree attraction is a way to diversify narrow rural economies and compensate for losses in extractive and manufacturing industries, some states have developed policies explicitly focused on attracting retirees. The chapter reviews a number of state-level retiree-attraction policies and explains the logic behind such schemes.

Chapter 3, "The Formation and Development of Rural Retirement Destinations," uses county-level census analysis to expand the community-level focus introduced in Chapter 2. Five hundred and eighty-four counties have been classified as rural retirement destinations at one period or another since 1980, but only 111 have been in this category continuously from 1980 to the present. Accordingly, the first part of this chapter focuses on movement into and out of the RRD category over a 20-year period. We examine the impact of metropolitan reclassification on the loss of RRD status because some rural retirement destinations leave the category when they are re-classified as metropolitan areas. We find that this factor is relatively unimportant during the 1980s, but it becomes increasingly important during 1990–2000. We then attempt to develop a comprehensive understanding of why some rural counties are more likely than others to become or maintain status as RRDs.

The second half of the chapter reports the results of a multivariate analysis of factors associated with being a RRD in 2000 and with becoming or losing RRD status between 1990 and 2000. We examine the impact of three domains of demographic, economic and geographic variables on these outcomes; but we find that only population change, percent aged 65 and above and increases therein, having a high dependence on recreation and tourism or having a low dependence on farming contribute to explaining why some counties are more likely than others to be or become RRDs. We also find that once a county becomes a RRD, it is more likely to be retained in the category if it is more highly urbanized and growing, if it is older and aging more rapidly and if its population is more highly educated. In contrast, RRDs are more likely to drop out of the category if they are becoming more racially diverse, have comparatively high in-migration from other states, are in the South and/or have a high natural amenity score. These analyses are a first step toward understanding why some places are more likely than others to be or become a RRD, but like most social science analyses they raise as many questions as they resolve. While a step in the right direction, our analyses fall short of accomplishing the goal of developing a comprehensive explanation of this phenomenon.

Chapters 4 and 5 focus on in-migrants themselves and report results from our panel survey. Chapter 4, "Who Moves to Rural Retirement Communities and Why Do They Move There?" begins by reporting that three-quarters of older in-migrants moved from metropolitan areas. We also found that in-migrants were about evenly split between persons who moved from another state and those who were intra-state migrants. The analysis then focuses on migrant selectivity and confirms the now well-accepted finding that compared with longer-term older residents, in-migrants are positively selected with respect to age, marital status, living arrangements and education. The chapter's second half examines migration decision-making among in-migrants to rural retirement destinations. This section is based on survey respondents' answers to open-ended questions about reasons for leaving previous residences and reasons for choosing to move to their current residence. If we can draw any conclusions from these open-ended responses it is that migration decision making involves multiple reasons. Perceived negative community attributes including urban externalities like traffic, crime or pollution and rural characteristics like smallness or remoteness were the most important reasons given for leaving one's previous residence and moving to a rural retirement destination. Other important "pushes" included life course changes such as retirement, divorce, or death of a spouse, poor weather and perceived lack of environmental amenities, lack of family members close by, high cost of living and taxes, and, somewhat surprisingly, a lack of employment opportunities. Similarly, about one quarter of respondents identified community attributes as their main reason for moving to a rural retirement community. Next in importance were having family nearby, and favorable environmental amenities.

Chapter 5 examines the social relationships in-migrants form after moving to rural retirement destinations. We investigate why some in-migrants are more likely to become socially integrated than others and explore the outcomes of social integration with respect to migrants' health and functional abilities. We were surprised to find that older in-migrants have little difficulty becoming socially involved in their new communities. In fact, their level of both formal and informal participation are quite comparable to that of longer-settled older persons in these same communities. Our multivariate analysis in the second part of the chapter provides modest support for the hypothesis that social participation benefits older persons' health. However, we find this only to be true of formal participation in clubs, organizations, volunteer activities and religious congregations. In contrast, relationships with family and friends have either no impact on older persons' health, or it is negative.

We return to the community level in Chapter 6 where we analyze information gained from the four case studies conducted in a subset of the 14 survey counties. Our interviews with elected and appointed officials, business leaders and service providers indicate that older in-migrants make important contributions in their new communities. Their volunteer participation is wide ranging and they are often the mainstays of service organizations, arts and cultural groups and church-based activities. They donate money, provide professional expertise and perform organizational tasks that help maintain and sustain the community. In contrast, our case studies identified a number of ways in which older in-migration displaces longer-settled persons. The persons we interviewed were enthusiastic and complimentary about older in-migrants' community contributions, but they also recognized these negative implications. Both pros and cons need to be weighed in a balanced way when considering the challenges and opportunities associated with older in-migration. We now turn to Chapter 2 which introduces the rural retirement destination concept and sets the stage for all that follows.

NOTES

1. While the interstate migration rate of persons aged 5–59 declined, the number of interstate migrants increased because the nation's total population increased. Among older persons, the migration rate remained steady at around 4.5 percent, and the number of migrants increased from 1.6 million to 2.1 million between the two periods 1975–80 and 1995–2000 (Longino, 2006).
2. Life expectancy at age 65 was 16.4 years for men and 19.4 years for women. Life expectancy increased for both sexes.
3. Moreover, the proportion of older persons with high income (400 percent or more of the poverty threshold) has also increased over this time. The highest quintile receives 19 percent from assets and 20 percent from pensions while the fourth quintile receives 10 percent from assets and 25 percent from pensions. In contrast, the lowest two quintiles receive 6 percent and 11 percent respectively from assets and pensions (U.S. Census Bureau, 2003b).
4. The poverty rate among older persons was 35 percent in 1959 (U.S. Census Bureau, 2003b).
5. Expressed in 2002 dollars.
6. The European Science Foundation maintains a bibliography of almost 400 citations dealing with retirement migration (European Science Foundation, 2002).
7. Due to sample attrition, the number of in-migrants on wave 2 was 301 and the number of longer-term residents was 337. We lost respondents because of death, refusal, out-of-date addresses and out-migration.

8. Findings in Chapter 2 indicate considerable overlap between RRDs and natural decrease counties (141 of 274 in 2000).

REFERENCES

American Association of Retired Persons. (2002). *Baby boomers envision their retirement: An AARP segmentation analysis.* Washington, D.C.: AARP.

Anderson, R., and Newman, J. (1973). Societal and individual determinants of medical care utilization in the United States. *Milbank Memorial Fund Quarterly, 51*, 95–124.

Atterton, J. (2006). *Aging and coastal communities.* Newcastle: Centre for Rural Economy, University of Newcastle upon Tyne.

Biggar, J. (1980). Who moved among the elderly, 1965–70: A comparison of types of older movers. *Research on Aging, 2*(1), 73–91.

Brown, D.L. (2002). Migration and community: Social networks in a multilevel world. *Rural Sociology, 67*(1), 1–23.

Brown, D.L., Fuguitt, G., Heaton, T., and Waseem, S. (1997). Continuities in size of place preferences in the United States, 1972–1992. *Rural Sociology, 62*(4), 408–428.

Champion, A., and Sheppard, J. (2006). Demographic change in rural England. In P. Lowe and L. Speakman (Eds.), *The ageing countryside: The growing older population of rural England* (pp. 29–50). London: Age Concern England.

Economic Research Service-USDA. (2007). Nonmetro America faces challenges from an aging population. *Briefing Room.* Retrieved March 2007, from www.ers.usda.gov/Briefing/Population/Challenges.htm.

European Science Foundation. (2002). *Bibliography on retirement migration.* Retrieved March 2007, from www.shef.ac.uk/sisa/esf/EW_Bibliography.shtml.

Federal Interagency Forum on Aging Related Statistics. (2004). *Older Americans 2004: Key indicators of well-being.* Washington D.C.: U.S. Government Printing Office.

Fuguitt, G., and Beale, C. (1993). The changing concentration of the older nonmetropolitan population 1960–1990. *Journal of Gerontology: Social Sciences, 48*(6), S278–S288.

Fuguitt, G., Beale, C., and Tordella, S. (2002). Recent trends in older population change and migration for nonmetro areas. *Rural America, 17*(3), 11–19.

Fuguitt, G., Brown, D.L., and Beale, C. (1989). *Rural and small town America.* New York: Russell Sage Foundation.

Fulton, J., Fuguitt, G., and Gibson, R. (1997). Recent changes in metropolitan-nonmetropolitan migration streams. *Rural Sociology, 62*(3), 363–384.

Garlick, S., Waterman, P., and Soar, J. (2006) Human capital, regional growth and productive ageing: New perspectives for policy and practice. In *Ageing-in-place: Implications for local government* (pp. 58–69). Australian Capital Territory: Australian Local Government Association.

Glasgow, N. (1980). The older nonmetropolitan migrant as a factor in rural population growth. In A. Sofranko and J. Williams (Eds.), *Rebirth of rural America: Rural migration in the Midwest* (pp. 153–170). Ames, Iowa: North Central Regional Center for Rural Development.

Glasgow, N. (1995). Retirement migration and the use of services in nonmetropolitan counties. *Rural Sociology, 60*(2), 224–243.

Glasgow, N., and Reeder, R. (1990). Economic and fiscal implications of non-metropolitan retirement migration. *The Journal of Applied Gerontology, 9*(4), 433–451.

Glasgow, N., and Sofranko, A. (1980). Migrant adjustment and integration in a new residence. In A. Sofranko and J. Williams (Eds.), *Rebirth of rural America: Rural migration in the Midwest* (pp. 87–104). Ames, Iowa: North Central Regional Center for Rural Development.

Haas, W., and Serow, W. (2002). The baby boom, amenity retirement migration, and retirement communities: Will the golden age of retirement continue? *Research on Aging, 24*(1), 150–164.

Illes, S. (2005). Elderly immigration to Hungary. *Migration Letters, 2*(2), 164–169.

Johnson, K., and Fuguitt, G. (2000). Continuity and change in rural migration patterns. *Rural Sociology, 65*(1), 27–49.

Johnson, K., Voss, P., Hammer, R., Fuguitt, G., and McNiven, S. (2005). Temporal and spatial variation in age-specific net migration. *Demography, 42*(4), 791–812.

Johnson, N. (2004). Spatial patterning of disabilities among adults. In N. Glasgow, L.W. Morton, and N. Johnson (Eds.), *Critical issues in rural health* (pp. 27–36). Ames, Iowa: Blackwell Publishing.

King, R., and Patterson, G. (1998). Diverse paths: The elderly in Tuscany. *International Journal of Population Geography, 4*(2), 157–182.

King, R., Warnes, A., and Williams, A. (2000). *Sunset Lives: British Retirement Migration to the Mediterranean.* New York: Berg.

Le Mesurier, N. (2006). The contributions of older people to rural community and citizenship. In P. Lowe and L. Speakman (Eds.), *The ageing countryside: The growing older population of rural England* (pp. 133–146). London: Age Concern England.

Lee, E. (1980). Migration of the aged. *Research on Aging, 2*(2), 131–135.

Lichter, D., Fuguitt, G., Heaton, T., and Clifford, W. (1981). Components of change in the residential concentration of the elderly population. *Journal of Gerontology, 36*(4), 480–489.

Litwak, E., and Longino, C. (1987). Migration patterns among the elderly: A developmental perspective. *Gerontologist, 27*(3), 266–272.

Longino, C. (1992). The forest and the trees: Micro-level considerations in the study of geographic mobility in old age. In A. Rogers (Ed.), *Elderly migration and population redistribution* (pp. 23–34). London: Bellhaven Press.

Longino, C. (2006). *Retirement migration in America.* Houston: Vacation Publications.

Longino, C., and Bradley, D. (2006). Internal and international migration. In R. Binstock and L. George (Eds.), *Handbook of aging and the social sciences.* (6th ed., Ch. 5). San Diego: Academic Press.

Marshall, V., Longino, C., Tucker, R., and Mullins, L. (1989). Health care utilization of Canadian snowbirds: An example of strategic planning. *Journal of Aging and Health, 1*(1), 150–168.

Morton, L.W. (2004). Spatial patterns of rural mortality. In N. Glasgow, L.W. Morton, and N. Johnson (Eds.), *Critical issues in rural health* (pp. 37–45). Ames, Iowa: Blackwell Publishing.

Nelson, P., Nicholson, J., and Stege, H. (2004). The baby boom and nonmetropolitan population change, 1975–1990. *Growth and Change, 35*(4), 525–544.

Pillemer, K., Moen, P., Wethington, E., and Glasgow, N. (Eds.). (2000). *Social integration in the second half of life.* Baltimore: Johns Hopkins University Press.

Plane, D. (1992). Age-composition change and the dynamics of interregional migration in the U.S. *Annals of the Association of American Geographers, 82*(1), 187–199.

Plane, D., and Rogerson, P. (1991). Tracking the baby boom, the baby bust, and the echo generations: How age composition regulates U.S. migration. *The Professional Geographer, 43*(4), 416–430.

Redwood, F. (2006, April 30). Bucolic bliss. *The Sunday Times*, 6.

Reeder, R. (1998). *Retiree attraction policies for rural America* (Agriculture Information Bulletin No. 741). Washington, D.C.: U.S. Department of Agriculture, Economic Research Service.

Rodriguez, V., Fernandez-Mayorales, G., and Rojo, F. (1998). European retirees on the Costa del Sol: A cross-national comparison. *International Journal of Population Geography, 4*(2), 183–200.

Rowles, G. and Watkins, J. (1993). Elderly migration and development in small communities. *Growth and Change*, 24(FALL), 509–538.

Sofranko, A., Fliegel, F., and Glasgow, N. (1983). Older urban migrants in rural settings: Problems and prospects. *International Journal of Aging and Human Development, 16*(4), 297–309.

Stallman, J., Deller, S., and Shields, M. (1999). The economic and fiscal impact of aging retirees on a small rural region. *The Gerontologist, 39*(5), 599–610.

United Nations Population Division. (2003). *World population prospects: The 2002 revision.* New York: United Nations.

U.S. Census Bureau. (2003). *Internal migration of the older population 1995 to 2000* (Census 2000 Special Reports-10). Washington D.C.: U.S. Department of Commerce.

U.S. Census Bureau. (2003b). *Current population survey, annual social and economic supplement, 1960–2003.* Washington, D.C.: U.S. Department of Commerce.

U.S. National Center for Health Statistics. (2000). *National nursing home survey.* Rockville, Md.: National Institutes of Health.

Wake Forest University. (2007). *Elderly migration bibliography.* Winston-Salem, N.C.: Reynolda Gerontology Program. Retreived March 2007, from www.wfu.edu/gerontology/ retirement-migration-biblio-info.htm.

Warnes, A. (2001). The international dispersal of pensioners from affluent countries. *International Journal of Population Geography, 7*(5), 373–388.

Weeks, J. (2005). *Population: An introduction to concepts and issues.* Belmont, C.A.: Wadsworth.

CHAPTER 2

RETIREMENT DESTINATIONS
IN THE COUNTRYSIDE

INTRODUCTION

In the 1970s, for the first time in over 100 years, the US rural population growth rate exceeded that in urban areas. This renewed rural population growth was primarily a result of net in-migration, and was widely spread throughout the nation's regions (Brown and Wardwell, 1981). However, even though this new rural growth was relatively widespread, it tended to concentrate in particular types of areas, including those that were attracting large numbers of retirement-age migrants. As Glasgow (1988) has shown, retirement destinations have been consistent rural growth nodes in each decade since the 1970s. In fact, rural retirement destinations continued to grow rapidly, and had net in-migration during the 1980s when most other rural county types of areas lost population to their urban counterparts (Reeder, 1998).

Johnson and Cromartie (2006) have demonstrated that older in-migration to rural areas has been a continuous and uninterrupted process during the past 40 years, and especially during the 1970s and 1990s (see Figure 2.1). During the 1950s, rural areas lost older migrants to urban areas, but at a much lower rate than the losses experienced at younger ages. Then, during the 1960s, while the total rural sector had net out-migration of over 0.5 percent per year, migration turned positive at ages 60 to 74. In other words, positive in-migration at ages 60 and older began a decade prior to the overall rural population turnaround and has persisted since then. The fact that rural areas have attracted older in-migrants during periods of relatively weak rural economic performance, when other age groups either stayed put or moved to urban areas, speaks to the fact that most older persons move to rural areas to accomplish life style, family, and amenity goals, not in search of jobs. Rural retirement destinations tend to be perceived as attractive places that are rich in environmental amenities and leisure activities. Knowing that older people have been moving to rural areas for almost 50 years poses a host of questions about the characteristics of the places they

Figure 2.1. Nonmetro Age-Specific Net Migration Rates, 1950–2000

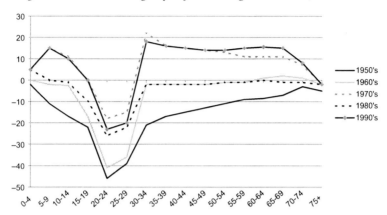

move to, the community-level impacts of in-migration by older persons, and the extent to which states and localities actively seek to recruit older in-migrants. These issues are addressed in this chapter.

As indicated in Chapter 1, most people do not migrate to a new community when they retire, but of those who do move, a disproportionate share find their way to rural areas (Beale, 2005; Longino and Bradley, 2003). Nevertheless, not all rural areas are equally likely to be retirement destinations. This chapter identifies the types of rural areas that have attracted older migrants during the last 40 years. First, we develop a conceptual framework for considering such areas, and then we use this frame to identify rural retirement destinations, describe their characteristics, and compare them with other types of rural places. Because many states and localities have concluded that recruiting retirees is an attractive economic development option, they have implemented explicit strategies to recruit retirees (Reeder, 1998). The chapter concludes by describing some of these recruitment strategies in the context of research that has examined the place-level impacts of retirement migration.

RURAL RETIREMENT DESTINATION COMMUNITIES DEFINED

When people think about American retirement communities they often visualize age-segregated, comprehensively planned, gated areas in Florida or Arizona such as *Sun City* or *Leisure World*. While *Sun City* and its ilk are an aspect of the retirement community phenomenon, retirement-based development takes place in many different types of places. This being acknowledged, it is useful to develop a conceptual framework for identifying retirement destinations that minimizes variability among places in the category and helps to differentiate them from other settlement types. Relatively high net in-migration at older ages, of course, is the defining characteristic of membership in the retirement destination category. However, it is important to acknowledge that in-migration at older ages takes place not only in planned communities such as *Sun City* and *Leisure World*, but in a wide variety of unplanned retirement destinations.

This distinction between planned and unplanned retirement destinations has shaped research and policy on the retirement migration issue. As Stephen Golant (2002) has observed, however, the conventional distinction between *planned elder retirement complexes* (PERC) and *naturally occurring retirement communities* (NORC) is not without problems. In fact, each category is so diverse that a simple comparison between them is of questionable value. PERCs share the common property of being residential areas especially built by private corporations or by nonprofit organizations to accommodate older persons, but the similarity ends there. The category includes options for frail older people, such as congregate housing and assisted living, and options for active older persons, such as age-segregated subdivisions of various sizes with a clubhouse and swimming pool. Also falling into this category are the large self-contained recreation-dense retirement villages that include residential, recreational and commercial facilities, and a full range of infrastructure and civic institutions. In addition, continuing care retirement communities that accommodate the transition from an active life to one where more assistance is needed are becoming more prevalent.

In contrast to PERCs, which share the fundamental characteristic of being planned, Golant believes that the NORC category has no unifying characteristic and is therefore fundamentally flawed. In his opinion, NORCs are not natural, not usually communities in the sociological sense, and many of their residents are not retired. To reduce this ambiguity, Golant proposes the term *deliberately occupied but unplanned elder residences* (DOUER). This category includes a range of areas where older persons have concentrated, but where the enclave of older people is not a planned development (Golant, 2002). Many DOUERs are located in the suburban peripheries of large American metropolises, but others are located in rural regions. Thus the DOUER concept, with its emphasis on unplanned but deliberate occupation by older residents, approximates but is not completely consistent with the rural retirement destinations we studied in this research. Because the DOUER concept is heavily focused on housing and the *elder residence* per se, it masks many of the broader community issues that concern us in this research. Our study has a community-wide focus inasmuch as we examine the destinations where older persons move, as well as the impact of residential relocation to such areas on older migrants' health and well-being. Accordingly, we employ the term *Unplanned Retirement Destination Community* (URDC). The essential characteristic of URDCs is net in-migration of people aged 60 and older, but some of these areas are also characterized by aging-in-place among longer term older residents. In other words, URDCs are often simultaneously occupied by longer-term older residents and by older in-migrants. URDCs generally include a range of socio-economic classes and hence contain a mix of wealthier and more modest neighborhoods, as is true of similar sized rural communities that lack significant rapid growth of the number of older residents. For ease of narrative in this book we refer to URDCs as *rural retirement destinations* (RRDs).

Our Definition of Rural Retirement Destinations

As indicated above, relatively high in-migration at older ages is the foundational element of the unplanned retirement destination settlement category. This raises two questions: (a) how old is *old*, and (b) how high is *high*? While distinctions like this are always somewhat arbitrary, we followed the U.S. Department of Agriculture's (USDA) convention by operationally defining rural retirement destinations as *nonmetropolitan counties with 15% or higher net in-migration at ages 60 and older* (ERS-USDA, 2004).[1] We will proceed to look at each element of this definition separately.

Even though 65 is often considered to be America's *retirement age*, research shows that many persons retire fully from the labor force earlier, and that some persons begin the transition to retirement by moving to less than full-time employment during their early 60s. In 2005, 42 percent of men aged 60–64 and 54 percent of women aged 60–64 were already out of the labor force (Gendell, 2006). Of those who were still working for pay, about 8 out of 10 men and 7 out of 10 women at these ages were working full time. Because age 60 includes a high percentage of both retirees and individual's phasing out of the workforce, it is a reasonable lower limit for measuring retirement migration.[2] Consistent with Gendell's observations, however, it should be kept in mind that while most of the older rural in-migrants studied in this book are retirees, some are full-time workers and some are transitioning from full-to part-time employment.

The choice of a migration rate is also somewhat arbitrary, but as Longino and Bradley (2003) have indicated, interstate migration among older persons has been rather stable since 1960 at about 4–5 percent per decade. Additionally, census data show that, during the 1990s, 4.2 percent of persons aged 65+ were inter-state migrants and 9.1 percent were inter-county migrants (He and Schacter, 2003).[3] Accordingly, the 15 percent lower limit for migration is significantly higher than the migration rate experienced by older persons in most counties, and identifies areas where older migration is of particular importance. Analysis of county level age-specific migration data indicate that the nonmetropolitan net migration rate at ages 60–69 was about 8 percent during the 1990s (Johnson,

Fuguitt, Hammer, Voss and McNiven, 2003). Hence, the 15% lower limit does a good job of identifying retirement migration destinations and distinguishing them from other rural places where older in-migration is less prevalent.

One may also object to the use of whole counties as the units of analysis in research focusing on local organization and change, and we do not disagree with many of the objections that have been raised by critics (Morrill, Cromartie and Hart, 1999). Counties have many benefits as units of analysis, including the fact that their boundaries are relatively stable over time and that a large amount of socio-economic and demographic data, including age-specific net migration rates, are available at the county level. Moreover, counties serve as a prime building block for the nation's system of statistical geography, and they raise revenue and provide essential services and functions. Therefore, even though counties may not be genuine communities in the sociological sense (Delanty, 2003), and while we understand that retirement communities are embedded within larger counties, we contend that much can be learned about the community-level aspects of rural retirement migration by examining the phenomenon at the county level. That being said, we do not believe that county-level analysis alone is a sufficient basis for examining the rural retirement migration phenomenon. As will be discussed in Chapters 3 through 6, we have employed a multi-level and multi-method research design that combines county-level census analysis, survey data collected from older persons living in rural retirement destination counties, and in-depth case studies conducted in four purposely selected rural retirement destinations.

Finally, since our focus is on *rural* retirement migration, we had to decide how best to delineate rural areas. The vast literature on the nature of rurality in highly developed societies like America features spirited debates between social constructivists (Halfacree, 1993, 2003) and structurialists (Brown and Cromartie, 2003), as well as debates within each of these camps. In this research we have chosen to use a more structuralist approach because it provides concrete criteria for locating and integrating our three data collection activities. By making this decision we do not disdain the more interpretive approach. As Brown and Kandel (2006, p. 10) have observed, "While the term *rural* has intuitive meaning replete with pastoral imagery, few definitions can fulfill most or all of the requirements of researchers, program managers, and policy analysts who rely upon them." In this research, we use the official residential classification scheme employed by the U.S government. While the U.S. currently employs two different parallel residential measures, "rural" and "nonmetropolitan," we chose nonmetropolitan to approximate rural in this study.[4] Since the nonmetropolitan category is comprised of entire counties, it has many of the advantages (and disadvantages) identified earlier when discussing our rationale for using counties in the community-level parts of our study. We believe it does a good job of delineating the geographic location of the social process we are examining, namely the movement of older persons into rural areas. For ease of exposition, we use the terms "rural" and "nonmetropolitan" interchangeably, but when these terms are used in connection with the analysis of county-level census data, the actual residential categorization employed is nonmetropolitan, not rural.

THE GEOGRAPHIC LOCATION OF RURAL RETIREMENT DESTINATIONS

The USDA definition of rural retirement destinations yields 274 counties.[5] Hence, only about 13 percent of all rural counties qualify as retirement destinations. The locations of these 274 counties are shown in Figure 2.2, and they are additionally listed in the Appendix to this chapter (Beale, 2005). Unsurprisingly, the majority of these areas are located in warm weather regions such as Florida, the Texas hill country, and the desert Southwest. Surprisingly, only three rural retirement destinations are located in California, and research by Longino and Bradley (2003) indicate that California was a net exporter of older persons in both the 1980s and the 1990s. In fact, after New York, California had the second highest gross out-migration at ages 60+ of any state and serves as an

Figure 2.2. Nonmetro Retirement Destination Counties, 2000

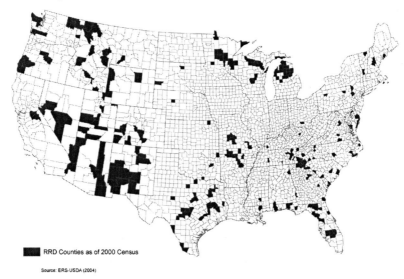

RRD Counties as of 2000 Census

Source: ERS-USDA (2004)

important source of older in-migration received by Arizona, Nevada and other southwestern states. What is surprising about the location of rural retirement destinations is that, while they are relatively concentrated in the South and Southwest, concentrations are also located in the Upper Great Lakes, the Ozark Mountains, southern Appalachia, the front range of the Rocky Mountains and the Pacific Northwest. It is a mistake, then, to think that rural retirement migration is a solely sunbelt phenomenon, as retirement destinations are widely scattered across the country. The photographs in Figures 2.3a,b,c,d show the diversity that exists within the RRD category.

RURAL RETIREMENT DESTINATIONS IN A COMPARATIVE PERSPECTIVE

Retirement Destinations Versus Natural Decrease Areas

Geographic Location

Population aging is a defining socio-demographic feature in two types of rural areas in the U.S: (a) retirement destinations, and (b) areas that have experienced natural decrease (i.e., more deaths than births). These two types of areas, however, have distinctly different short- and longer-term patterns of population growth, economic activities and prospects for future growth, development and well-being. In contrast, their demographic and socio-economic compositions are more similar than typically realized and, as will be shown shortly, there is significant overlap in their patterns of geographic location.

The map displayed in Figure 2.4 shows that natural decrease counties are concentrated in the nation's mid section, especially in the Northern Great Plains and Western Corn Belt.[6] The concentration of natural decrease in the Northern Great Plains is closely related to high dependence on farming in these areas. Nearly 60 percent of farming dependent counties—those which derive at least 20 percent of labor and proprietors' income from agriculture—had more deaths than births between 1999 and 2000. For most of these areas, natural decrease is a persistent condition that first appeared as early as the 1970s (Johnson, 2004).

Retirement destinations, in contrast, are more widely spread and tend to locate in the vicinity of attractive natural amenities such as lakes, coasts, mountains and forests. Warm weather is a prominent factor in retirement migration, but not a necessary determinant. Indeed, retirement destinations are found in the Upper Great Lakes, in New England and in the Northern Rocky Mountains, areas known for snow and cold temperatures rather than for sunshine and warm weather.

Figure 2.3a. Gila County, Arizona (Payson)

Figure 2.3b. Leelanau County, Michigan (Suttons Bay)

Figure 2.3c. Lincoln County, Maine (New Harbor)

Figure 2.3d. Transylvania County, North Carolina (Brevard)

Figure 2.4. Natural Decrease During 2000–2004

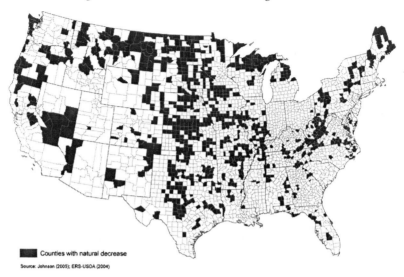

Counties with natural decrease

Source: Johnson (2005); ERS-USDA (2004)

The Overlap of Natural Decrease and Retirement Age In-Migration

The overall geographic patterns of natural decrease and retirement in-migration, while markedly different, masks the fact that 141 of 274 RRD counties are also natural decrease counties (Figure 2.5). These overlapping counties are concentrated in two midwestern (Michigan, Wisconsin), and three southern states (Georgia, North Carolina and, especially, Texas). None of these states fit the general image of natural decrease, e.g., persistent out-migration of young adults, chronic population decline, heavy dependence on farming, and location in the Great Plains; yet as will be shown later in this chapter and discussed in Chapter 3, long term *in-movement* at older ages can also produce an age structure that produces fewer births than deaths. What is surprising, however, is that even though natural decrease counties that are also retirement destinations have high percentages aged 65 and older, they are not demographically stagnant like most other natural decrease counties. In fact, as shown in Table 2.1, these RRD/ND overlap counties grew by 21 percent between 1990 and 2000 and had an overall net migration rate of 23.1 percent. Both of these figures far exceed those experienced in natural decrease counties that are not retirement destinations.

Figure 2.5. Nonmetro Counties With Both Natural Decrease and Retirement In-Migration And Other Natural Decrease Counties

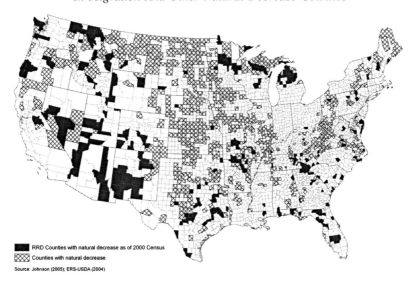

RRD Counties with natural decrease as of 2000 Census
Counties with natural decrease
Source: Johnson (2005); ERS-USDA (2004)

Demographic Dynamics

The demographic dynamics that produce natural decrease are considerably different from the patterns of population change occurring in retirement destinations. Natural decrease is a result of chronic out-migration of persons in their 20s and 30s which produces an age structure bereft of persons in their child bearing ages. Accordingly, even if remaining parents in these counties bear children at average or higher rates, these areas cannot produce very many births because so few potential parents are available. Johnson and Rathge (2006) have shown that only 18,123 children were born in the 1418 farming-dependent Great Plains counties in 2000. As a result, the population ages to the point where more people die than are born. In the absence of net in-migration, these areas are demographically stagnant, and can expect to stay on a downward escalator of continuous stagnation or population decline in the future.

Most retirement destination counties, in contrast, not only attract older migrants but also working age persons and persons in the child bearing years. This is because retirement migration induces economic growth and employment gains, thus retaining and attracting employers and employees. Ironically, net in-migration of older persons in retirement destination counties does *not* necessarily contribute to extreme population aging. As shown in Table 2.2, the percentage of

Table 2.1. Population Dynamics in Rural National Decrease Counties, 1990–2000

	Natural Decrease[a]	Natural Decrease Not RRD	Natural Decrease[a] and RRD[b]		
			Total	Continuous 1979–2004	Not Continuous
Number of Counties	837	696	141	68	73
Percent 65+	19.1	18.9	19.9	21.4	18.6
Percent population change 1990–2000	3.6	−0.01	21.1	25.4	17.2
Percent lived in different county in 1995	17.7	16.6	23.1	24.9	21.5
Mean Population size in 2000	16907	15376	24427	28686	20528

Source: U.S. Census of Population, 1990 and 2000.
[a] Natural Decrease provided by Ken Johnson.
[b] Retirement Destinations as defined by USDA-ERS based on results of 2000 census.

the population aged 65 and older in retirement destinations held steady at a little under 18 percent between 1990 and 2000. While this is two percent higher than in other rural counties, the percentage aged 20–64 is the same in both categories, and the percent under age 20 is only slightly lower in retirement destinations.[7] Moreover, retirement destination counties had an overall net migration rate of 22.9 percent, and their rate of population growth was 25.2 percent between 1990 and 2000. In contrast, farming-dependent counties in the Great Plains had

Table 2.2. Age Composition of Rural Retirement Destinations in Comparison With Other Nonmetro Counties

	1990			2000		
	RRD	Natural Decrease	Other Nonmetro	RRD	Natural Decrease	Other Nonmetro
Percent 0–19	27.4	27.8	30.3	26.4	26.1	28.4
Percent 20–64	54.7	53.0	54.1	56.4	54.8	55.8
Percent 65+	17.9	19.2	15.7	17.2	19.1	15.8

Source: U.S. Census of Population, 1990 and 2000.

net out-migration of −2.5 percent, and an overall population loss (Johnson and Rathge, 2006).

Rural Retirement Destinations Compared With Other Types of Rural Areas

In 1985, the USDA's Economic Research Service (ERS) developed a policy-oriented classification of rural counties (Ross and Green, 1985). The classification system, generally referred to as the "ERS typology," provides an identification of groups of nonmetropolitan counties which share important social, economic and policy-relevant characteristics. The original typology, created in 1980, summarized the diversity of rural economic and social conditions among nonmetropolitan counties into eight overlapping types. Five of the types reflected dependence on particular economic activities including farming, manufacturing, mining, government and a non-specialized residual category, while the remaining categories involved three policy relevant themes including persistent poverty, federal government ownership of a significant proportion of the county's land and in-migration of retirement age population.

The typology has been updated several times to reflect changing rural conditions (Cook and Mizer, 1994; ERS/USDA, 2004). The typology's most recent revision includes six non-overlapping economic types which show employment dependence on various sectors of the economy (farming, manufacturing, services, government, mining and non-specialized) plus seven overlapping policy-types showing the location of various aspects of development or underdevelopment (chronic population loss, the presence of a developed recreation and tourism industry, high prevalence of housing stress, low educational attainment, low employment and retirement migration destinations). The expansion of the typology's policy-related categories reflects the increased complexity of social and economic life in contemporary rural America, and the recognition that rural policy is a multi-faceted enterprise.

That the retirement destination category has been part of the ERS county typology throughout its various revisions shows the continuing importance of this type of settlement. Of course, the retirement destination category is not necessarily a pure type, as it may overlap with other types of areas. Accordingly, the data in Table 2.3 show the overlap between being a retirement destination and other county-level economic specializations or policy themes. The general expectation would be that retirement destinations share geographic space with the service and recreation/tourism dependent county types. In contrast, because retirement migration is typically thought to promote population growth and economic well-being, one would expect little or no overlap with persistent poverty (even though both poverty and retirement are overrepresented in the South), population loss and heavy dependence on farming.

The data in Table 2.3 support most of these expectations. Most importantly, almost one half of rural retirement destinations are also tourism and
recreation areas, which supports the observation that older persons are attracted
to areas with developed recreation and tourism industries. In contrast, only one of
five retirement destinations also have a high dependence on service employment.
While older persons require many services, older in-migration does not seem to
translate into an exceptionally high representation of services in local economies
where retirees congregate. As expected, only 5.8 percent of retirement destinations are farming-dependent areas; only 5 of the 274 retirement destinations
experienced recent population loss; and only 10 percent have extremely high and
persistent poverty rates. The most unexpected overlap between retirement destinations and another category was with counties that experienced natural decrease.
Little overlap was expected because, while retirement destinations are growing
relatively rapidly, natural decrease counties are stagnant at best. Nonetheless,
the data show that 51% of rural retirement destinations are also natural decrease
locations. As shown previously in Figure 2.4, this coincidence between RRD
counties and natural decrease occurs mostly in 5 midwestern and southern states
and is almost totally absent in the Great Plains, where the conventional image of
natural decrease as occurring in the midst of agricultural dependence pertains. It
should be remembered that these overlapping counties account for a little more
than half of retirement destinations, but only comprise 17 percent of all natural
decrease counties. As will be discussed in Chapter 3, these RRD/natural decrease
overlaps show that natural decrease can be the result of long term in-migration
of older persons as well as of the conventional scenario of chronic out-migration
of persons in the child bearing ages.

Table 2.3. Overlap Between Rural Retirement Destinations and Other County Types

	RRD Counties that Overlap Other County Types[a]	
	Number	%
Services dependent	56	20.2
Farming dependent	16	5.8
Recreation dependent	137	49.5
Population loss	5	1.8
Persistent poverty	28	10.1
Natural decrease[b]	141	51.3

Source: Johnson & Cromartie (2006).
[a] County types from USDA-ERS.
[b] Kenneth Johnson provided the authors with county codes to identify natural decrease
areas in March 2006 along with his permission to use them in this analysis. The natural
decrease counties used herein are the same as used by Johnson and Cromartie (2006).

The most important conclusions to be made from the overlap shown in Table 2.3 are that rural retirement destinations tend to be demographically dynamic and economically well off. In addition, they share geographic space with the burgeoning rural tourism and recreation industry. Rural retirement destinations are often rich in scenic beauty, including lakes, rivers, ocean shorelines, mountains and the outdoor recreation opportunities found in such areas. Golf, tennis, fishing and beach activities are typical, but so are skiing, snowmobiling and hiking in colder weather and mountainous areas. Less appreciated is that half of rural retirement destinations had more deaths than births, even though as shown in Table 2.1, these overlap counties are growing and attracting in-migration. This underlines important differentiation within the RRD category. While some rural retirement destinations are demographically dynamic with respect to all of the components of population change, others have in-migration but still experience more deaths than births.

Comparative Profile of Retirement Destination Counties

Now that we know where retirement destination counties are located, and how their location compares with that of other rural county types, we need to know something about their social and economic characteristics, e.g., what kinds of places are they, and how do they compare with other types of rural areas? This information will provide insight into the place character-istics that attract older migrants. The data in Table 2.4 provide a profile of retirement destination counties in comparison with other nonmetropolitan counties in both 1990 and 2000. Compared with other rural counties, rural retirement destinations are slightly older, have somewhat higher personal and household income, slightly lower poverty rates, and slightly higher educa-tional attainment. Nevertheless, these differences are modest and the data do not reveal important differences between retirement destinations and other rural counties with respect to marital status, homeownership or employment. The most striking differences between retirement destinations and other rural counties are reflected in the data on in-migration and population growth, both of which are markedly higher in retirement destinations. It is also notable that RRDs were larger than other nonmetropolitan counties in both 1990 and in 2000. The data in Table 2.4 also show that the percentage of RRDs that are adjacent to metropolitan areas exceeds that of non-RRD counties, and this difference increased substantially during the decade of the 1990s. This suggests that many of the new RRD counties are adjacent to metropolitan areas.

The data in Table 2.5 compare the characteristics of rural retirement destinations and other types of rural counties with which they share little

Table 2.4. Social and Economic Profile of Rural Retirement Destinations in Comparison of Non-Retirement Rural Counties, 1990 and 2000

	1990		2000	
	RRD	Not-RRD	RRD	Not-RRD
Percent 65+	17.9	15.7	17.2	15.8
Percent married, spouse present	48.8	46.1	46.2	44.1
Percent completed some college	36.9	32.3	43.0	39.5
Percent owner occupied house	74.6	72.8	76.2	73.1
Per capita income ($)	11266	10187	17263	16008
Median household income ($)	22569	21643	33203	31838
Percent poverty	16.1	18.0	13.9	15.2
Percent employed	51.8	55.3	52.4	56.0
Percent unemployed	3.9	4.0	3.6	3.5
Percent migration[a]	27.1	18.2	22.9	17.1
Percent population change[b]	22.1	−1.8	25.2	4.7
Mean population size	26881	21708	28360	23091
Percent adjacent to SMSA[c]	49.5	42.7	61.7	50.2

Source: U.S. Census of Population, 1990 and 2000.
[a] Percent of population living in another county (either in same or different state) 5 years prior to census.
[b] 1980–90 and 1990–2000;
[c] SMSA: Standard Metropolitan Statistical Area.

or no geographic space—farming-dependent, population loss and persistent poverty counties, and with natural decrease counties with which they overlap. Natural decrease counties have a somewhat older age structure than retirement destinations, but the difference in percentage aged 65+ is not as great as expected (19.1 percent vs. 17.2 percent).[8] Similarly, other socio-economic and demographic comparisons reveal only slight differences. Retirement destinations have somewhat higher college educational attainment and income, especially compared with persistent poverty counties. Likewise, both unemployment and poverty are distinctly higher in persistent poverty counties than in the other county types. As would be expected, retirement destinations have much higher levels of in-migration and population growth than any of the other areas.

Table 2.5. Social and Economic Profile of Rural Retirement Destinations in Comparison with Selected other Types of Rural Counties, 2000

	RRD	Natural Decrease	Farm Dependent	Population Loss	Persistent Poverty
Percent 65+	17.2	19.1	18.2	18.8	14.1
Percent married, spouse present	46.2	46.7	46.3	46.0	38.6
Percent completed some college	43.0	40.2	42.3	41.0	31.5
Percent owner occupied house	76.2	75.5	73.0	74.4	70.1
Per capita income ($)	17263	16342	15366	15758	12900
Mdn. household income ($)	33203	30775	30197	30072	24576
Percent poverty	13.9	13.9	15.6	15.2	25.7
Percent employed	52.4	54.3	57.4	56.2	47.9
Percent unemployed	3.6	3.1	2.6	3.0	4.8
Percent migration[a]	22.9	17.9	17.5	15.3	14.6
Percent population change	25.2	3.6	0.2	−5.8	6.1
Mean population size	28360	16805	7449	14506	19581
Percent adjacent to SMSA[b]	61.7	42.6	28.0	30.3	46.8

Source: U.S. Census of Population, 2000.
[a] Percent of population living in another county (either in same or different state) 5 years prior to census;
[b] SMSA: Standard Metropolitan Statistical Area.

It is interesting to note that while natural decrease and farming-dependent counties are demographically stagnant, they are not disadvantaged with respect to poverty, unemployment and other socio-economic measures. In other words, this comparison shows that while natural decrease areas are losing population, they are not poor. A similar picture is presented by comparing farming-dependent vs. retirement destination counties. Farming-dependent counties tend to be somewhat less well off than either rural retirement or natural decrease areas, especially as indicated by poverty rates (15.6 percent vs. 13.9 percent for both in 1999), and the total lack of population growth. The higher poverty in these areas is probably due to the fact that some of the farming-dependent counties are located in the Mississippi Delta. In general, however, farming-dependent areas are concentrated in the Midwest and Great Plains, where decades of out-migration has altered the age structure and contributed to demographic stagnation, but not necessarily to economic decline. One additional point is that farming-dependent areas have small sized populations. They contain only an average of around 7500 residents compared with over 28,000 in retirement destinations. Even counties characterized by natural decrease and population decline are considerably larger than farming-dependent areas.

The final comparison in Table 2.5 focuses on persistent poverty areas. Compared with rural retirement destinations, these data show that persistent poverty counties have all of the expected differences in demographic and socio-economic characteristics. They are substantially younger and have much lower rates of marriage (38.6 percent vs. 46.2 percent), college education attainment, income and home ownership. As expected both their poverty rate (25.7 percent) and unemployment rate are much higher than in any other county type. Migration and population growth are also lower in these areas compared with rural retirement destinations.

To summarize, the above comparisons show that rural retirement destinations are more demographically dynamic than other types of rural counties, especially as indicated by their much higher rates of in-migration. Rural retirement destinations attract older in-migrants, but also persons in the working and child bearing ages. These data show that rural retirement destinations are better off than most other types of rural areas; but, with the exception of persistent poverty areas, the differences are systematic but not large.

COMMUNITY-LEVEL IMPACTS OF RURAL RETIREMENT MIGRATION

Effects on Demographic and Socio-Economic Composition

Population Aging

First, it should be reiterated that retirement migration does not typically contribute to extreme population aging. As explained above, this is because places that attract older in-migrants also attract younger persons. Accordingly, while retirement destinations are slightly older than other rural areas, the difference is not great. Moreover, because in-migration to retirement destinations induces migration at younger ages, retirement migration leads both directly and indirectly to higher levels of overall population growth than would be expected if in-migrating retirees were the only contributor to growth (Reeder and Glasgow, 1990). This is because the younger in-migrants who are attracted by new jobs in services, retail, construction, and other industries tend to have larger households than older in-migrants. Additionally, childbearing age migrants are generally accompanied by their children and/or have children once they arrive.

Socio-Economic Status

A substantial amount of prior research demonstrates that in-migrating retirees are positively selected with respect to educational attainment and income status (Glasgow and Brown, 2006; Haas and Serow, 1993). Accordingly, retirement migration tends to enhance the destination community's socio-economic status. In addition, older in-migrants also bring experience and know-how to their new communities; and they are at least potentially available to contribute leadership and skilled volunteer labor to community organizations and activities (Le Mesurier, 2006). We will examine both of these issues in the context of rural retirement destinations in Chapters 4, 5 and 6 when we analyze survey and case study data on the socio-economic status and community participation of older migrants and longer-term older residents of rural retirement destinations.

Economic Impacts

Fagin and Longino (1993) examined income flows associated with the interstate migration of older persons and determined that Florida, Arizona, Texas and seven other states gained in excess of 100 million dollars annually during the 1980s while New York, Illinois, Ohio and seven other states lost over 100 million dollars. Clearly, retirement migration moves a lot of money around. Research, however, also shows that measuring the economic impacts of retirement migration is an exceedingly complex task. Studies have demonstrated that the economic effects of older in-migration involve both consumption and the employment growth that such consumption may induce. Moreover, it has been shown that direct consumption by older in-migrants should be differentiated from the indirect expenditures made by the public sector in providing goods and services to a growing population of older residents. In other words, analyzing economic impacts involves a consideration of jobs and income, as well as an examination of fiscal effects experienced at various levels of government (Baker and Speakman, 2006; Glasgow and Reeder, 1990; Stallman, Deller and Shields, 1999). The economic impact depends on the selectivity of older in-migrants and the programmatic responses of local governments in the newcomers' destinations. That being said, the preponderance of research has found positive economic impacts for receiving communities from the in-migration of older people (Serow, 2003; Serow and Haas, 1992; Stallman, Deller and Shields, 1999). Serow (2003) commented that there is remarkable consistency in the costs and benefits demonstrated by economic impact studies. These studies have found that retiree in-migration induces about 0.5 jobs per migrant; that public expenditure increases per migrant are in the range of $35–$40,000 per year; and that older in-migrants pay more in taxes than they cost in additional services.

The most sophisticated approach to measuring the economic impacts of older in-migration was developed by Judith Stallman and her colleagues (1999). Stallman and her research team developed a holistic simulation model of the economic and fiscal impacts of older in-migration, and applied it to a small rural economy in Wisconsin.[9] Their study showed that total earnings and total incomes increased as did retail sales. They also showed that government expenditures declined in some areas, suggesting the effects of growing economies of scale; they increased in others, reflecting the added capacity that had to be supplied to serve the growing older population. They also concluded that an increasing older population had a positive net fiscal impact because, while total non-educational expenditures increased, non-educational revenues grew even more rapidly.

Older in-migrants have a positive impact on jobs and income because their incomes are less vulnerable to down cycles in the national economy than is true of younger age groups that depend more heavily on wage and salary employment for their livelihoods. As Fagin and Longino (1993) have observed, income from investments fluctuates, but income from transfers and pension benefits is relatively stable. In comparing the recruitment of retirees with industrial recruitment, Fagin and Longino (1993) promote retiree attraction as a low risk way to diversify and stabilize a local economy. They contend that retirees increase the tax base, do not require large amounts of new infrastructure, do not demand subsidies or tax abatements, do not pollute the environment in the ways industry might, and are a pool of ready volunteers. Moreover, older in-migrants are relatively unlikely to compete for jobs with younger workers, while their consumption behavior is likely to induce the demand for additional workers in services, retail, finance, insurance, recreation, utilities, construction and real estate. It should be noted, however, that a recent study by Gendell (2006) showed that labor force participation among older persons has increased since 1994, which may eventually diminish the employment generating effect of older in-migration. And, as will be shown in Chapter 4, over one third of older in-migrants to the 14 RRDs in this study are at work, thus sometimes competing directly with younger workers.

Florida, Arizona and several other states have conducted evaluations of their retiree recruitment strategies and found that retirement attraction has large positive economic impacts. The Florida study (Destination Florida Commission, 2003, p. 2) for example, concluded that "direct spending by older Floridians and the value of their federal health benefits is estimated at $150 billion." This study also reported that "Florida's elderly residents represented a net benefit of $2.8 billion in taxes to state and local governments in the year 2000" (Destination Florida Commission, 2003, p. 2). The Arizona study (Rex and

Seidman, 2002) similarly concluded that retirees presented a clear, positive contribution to the state's private sector and that retirees were low users of public services. Moreover, the study noted that although retirees were low users of public services, the same is not true of younger persons who are attracted to the state to work in service industries catering to older persons. This underlines the complex nature of direct and indirect impacts that must be considered to judge the validity of retiree recruitment as an economic development strategy.

One specific economic concern that has been raised about retirement migration is whether the economic contribution made by older in-migrants might diminish over time as they age-in-place in their new communities (Serow, 2003; Skelley, 2004). Some scholars and policy makers anticipate that older in-migrants' net positive contributions to local economies will decline as they become older, suffer debilitating illnesses, experience the death of their spouse, lose other informal social supports, and/or cease driving. The concern is that older in-migrants will cease making positive contributions and eventually cost more than they contribute. Stallman and her colleagues (1999) considered this issue by running their economic and fiscal simulation model separately for young-old and old-old in-migrants. Their analysis showed that young-old retirees (aged 65–74) make significant positive contributions to the rural Wisconsin counties where they tested their model, and that older-old in-migrants (aged 75+) also have positive economic effects. Accordingly, they concluded that, "...increased local government expenditures are covered by the increased revenues, even as retirees age." (Stallman, Deller and Shields, 1999, p. 599) While Stallman's simulations provide some basis for optimism, we agree with Serow (2003) who contends that very little direct research has studied the economic and fiscal impacts of aging-in-place. Hence, the time horizon of costs and benefits associated with older in-migration should be considered when strategies for retiree recruitment are being considered.

ATTRACTING RETIREES AS A LOCAL DEVELOPMENT STRATEGY

At least 10 states currently have retiree attraction programs, and others are in the process of implementing similar strategies (Hass and Serow, 2002; Reeder, 1998). Retiree attraction policies focus on the young-old population (Skelley, 2004) who are typically married, in good health and economically secure. Retiree attraction is seen as a way to diversify narrow rural economies and/or to compensate for losses in extractive and manufacturing industries that are a result of long term economic restructuring and more recent processes of global off-shoring (McGranahan, 2003; Vias and Nelson, 2006). While research does not provide a definitive answer to the complex question of the costs and

benefits of retirement attraction, or how these costs and benefits change over
time as older in-migrants age-in-place, many rural regions of the U.S. have so
few options that they are willing to take a chance on retiree attraction schemes.
Moreover, this strategy has taken on added legitimacy because it has been
promoted by respected researchers who contend that it is "low risk," and superior
to industrial recruitment in terms of its direct and indirect economic stimulation
(Fagin and Longino, 1993).

Examples of State Programs to Recruit Retirees

In 1987, **Washington** established the *Art of Retirement* project which
was one of the first systematic state-wide efforts to attract retirees as an
economic development option. The state published a comprehensive self-help
guide instructing communities how to organize retiree recruitment campaigns
(Severinghaus, 1990). Since the 1980s, many Washington communities have
used the self-help approach to recruit older persons to rural communities. The
case of Chelewah, a town of 2000 persons located about 40 miles from Spokane,
garnered national attention in the early 1990s, claiming to have successfully
attracted 150 new residents in one year using the state's self-help techniques
(Reeder, 1998).

Alabama was also an early adopter of retiree-driven rural economic
development. Beginning in the late 1980s, the state's Department of Economic
and Community Affairs established the *Alabama Advantage for Retirees*
program. The Alabama model provides comprehensive assistance to rural
communities including planning and technical assistance, literature development,
marketing, finance, and amenity development (Reeder, 1998). The state initiated
an aggressive advertising program and developed a cooperative marketing
campaign so that groups of communities could join together and share the costs
of their recruitment activities. The state's most visible and effective action was
the creation of the $120 million *Robert Trent Jones Golf Trail* which attracts
golfers to the state, many of whom are retirees, pre-retirees and/or potential
retiree migrants. The state has also promoted its excellent fishing opportunities
in TVA reservoirs for both tourism promotion and retiree recruitment.

South Carolina's approach to retirement-led economic development
has been to focus on privately-developed, planned retirement communities.
Developers build and market the communities, while the state and local governments provide incentives such as free land and other subsidies (Reeder, 1998).
The state has held a number of conferences explicitly encouraging localities
to promote themselves as retirement destinations and publishes a magazine
promoting the state to potential retiree in-migrants. In the late 1990s, South
Carolina entered into negotiations with the Del Webb Corporation, the developer

of *Sun City* in Arizona, to build a new planned retirement community near Hilton Head.

North Dakota is hardly a retirement mecca, but in 1994 the state began operating *Project Back Home*, a comprehensive effort to attract former residents back to the state. Although not an official project of state government, current and former state economic development officials have promoted the idea. The program is a cooperative effort of the participating communities, the North Dakota Rural Electric Cooperative, and the North Dakota Central Data Cooperative (Reeder, 1998). Using direct mail, communities contact high school graduates who have moved from the state over the past 40 years. Persons who indicate a potential interest in returning to the state are then queried about their reasons for considering returning, and their community preferences and life style aspirations. Responses are matched with various communities where potential migrants' preferences can be met. This is an innovative and ingenious program, but no research has been conducted to determine its effectiveness in attracting migrants to return to the state.

Other states including Mississippi, North Carolina, Georgia, Arkansas and Michigan have also engaged in systematic efforts to attract retirees. All of these programs involve state marketing programs focused on the mature market, while some states also provide technical assistance to communities, community self-help manuals and various types of subsidies to private developers. Georgia, Mississippi and Michigan have enacted special legislation to reduce or eliminate taxation on retiree income (Reeder, 1998). In addition, California, Pennsylvania, New York, New Jersey and Ohio have sponsored special studies and conferences focused on the economic benefits of retiree attraction and/or retention, with the goal of stemming the flow of retirees to other states. Complementing these state-wide programs, many individual communities have mounted their own retiree recruitment efforts, though few of these programs have been systematically evaluated.

CONCLUSIONS

In this chapter we have conceptualized rural retirement destinations as unplanned rural retirement destination communities (URDC). This conceptualization indicates that we are focusing on entire rural communities where a substantially higher than average rate of older persons have chosen to migrate and not simply on older enclaves located within these larger areas. In addition, our work does not focus on planned, recreation-rich, gated communities such as *Sun City* or *Leisure World*.

Using this concept we demonstrated that retirement destinations are widely spread throughout the nation. While the majority are in the South and

Southwest, as would be expected, substantial numbers are in other regions not typically associated with "fun in the sun." Since many older persons migrate in search of amenities and leisure possibilities, their wide geographic dispersion indicates that the concept of amenities is inherently subjective. One person's preference may be another's worst nightmare. Accordingly, it is not only year round sunshine and sandy beaches, but cold winters, lots of snow and possibilities for snowmobiling, skating and cross country skiing that motivate the amenity-driven migration dynamic. Both situations can be thought of as amenity attributes that motivate certain older persons to move to rural retirement destinations.

We next asked what social and economic attributes characterize rural retirement destinations and differentiate them from other types of areas. Compared with other types of rural areas, we showed that retirement destinations are more demographically dynamic as indicated by their relatively high rates of net in-migration and population growth. In fact, rural retirement destinations have grown in each of the last four decades, regardless of the overall growth situation experienced by other types of rural areas.

The comparison between rural retirement destinations and rural areas experiencing natural decrease is particularly interesting. Both types of areas are characterized by older than average populations; but natural decrease areas are usually demographically stagnant while rural retirement destinations are growing much more rapidly than other rural locales. However, even though natural decrease areas are demographically sluggish, they are not socio-economically depressed. There is simply very little economic activity happening in most of the areas that are experiencing natural decrease. Because their age structure is so distorted by chronic out-migration of younger persons, they are on a demographic escalator going down while retirement destinations, even those with natural decrease, are advancing in the other direction. This picture is drawn from a comparison of the average experience of each county group, but of course all social categories have variability. As indicated above, only 17 percent of natural decrease counties are RRDs, but fully 51 percent of RRDs also experienced natural decrease. These overlapping counties have natural decrease because of long term in-migration of older persons, not because of chronic out-migration of youth. As shown in Figure 2.5, these RRD/ND counties are not farming dependent nor are they located in the Great Plains. In addition, as will be shown in Chapter 3, their rates of population growth and net migration at all ages far exceed those of other natural decrease counties. Accordingly, even though they have older age structures, they are not demographically stagnant places as is true of natural decrease counties which lack in-migration at older ages.

We next compared retirement destinations with other types of rural areas, and in each instance found that they were growing more rapidly and were socio-economically better off. Differences were especially great in comparison with persistent poverty counties even though both of these county types are concentrated in the South. Clearly, retirement in-migration is associated with positive population growth and above average rates of economic development. While our analysis does not establish whether retirement migration is a causal factor that results in enhanced community well-being, these two processes occur together.

Retiree recruitment is a complex strategy, and the jury is still out on how well it pays off or, indeed, if it pays off at all. The most sophisticated and comprehensive study of the economic and fiscal costs and benefits of retirement migration—that conducted by Stallman and her colleagues (1999)—was only tested in one small rural region of Wisconsin. Hence, one might question how generalizable Stallman's conclusions are to Alabama, Washington or even to Michigan, Wisconsin's upper midwestern neighbor. Other studies upon which retiree attraction strategies are based are often ad hoc considerations of one or another set of economic or fiscal indicators with no underlying model of how local economies grow and change. Moreover, none of the studies has used a longitudinal design to examine the time horizons of various costs and benefits as older in-migrants age-in-place. We thus recommend that states and localities proceed cautiously when considering this development option. In addition, retiree attraction is only one part of community economic development. As Skelley (2004, p. 214) has pointed out, "... communities must determine for themselves if a retiree attraction policy fits with their notion of a broader, sustainable development policy." Some communities embark on retiree attraction programs as if they were trying to attract any other industry; but, as Skelley (2004) has observed, retiree attraction strategies are substantively different from more general industrial attraction policies and require a different planning process to succeed. The direct and indirect effects must be considered, and they must be considered over a substantial time horizon. Moreover, attracting retirees as a development strategy concerns more than people, jobs and income. Far reaching impacts on institutions, civil society, and social relationships should also be considered. Our general feeling is that retiree recruitment is a sound policy, but to coin an old cliché, "more research is needed."

NOTES

1. According to USDA-ERS a county is a RRD if it grew by 15% or higher through in-migration at ages 60+ between 2 censuses. This was determined by computing the net migration rate for counties at ages 60+ using the forward survival rate method. First the population age 60+ is estimated, and then the enumerated population at 60+ is subtracted. The difference is net migration at age 60+. The estimated population

was attained by surviving the population aged 50–59 forward from 1990–2000 using a survival rate (based on a life table). This is aging-in-place and is conceptually the same as "births" into age 60+ category. Next deaths to persons age 60+ in 1990 (from vital statistics) are subtracted. In other words, this procedure is equivalent to determining natural increase at ages 60+ for a particular county. The difference between this estimated population age 60+ and the enumerated population age 60+ is net migration. If it is 15% or higher, the county is a RRD. See Voss, McNiven, Hammer, Johnson and Fuguitt (2004) for a discussion of the methodology used in this procedure.

2. However, recent research by Gendell (2006) demonstrates that labor force participation among persons age 60–64 has begun to increase in the US.

3. These mobility rates are much lower than those among younger persons, where we find that 20.0 percent moved across county lines, while 9.3 percent moved across state lines.

4. Nonmetropolitan counties were identified according to the metropolitan/nonmetropolitan delineation announced by the U.S. Office of Management and Budget in 2003 (U.S. Office of Management and Budget, 2003).

5. The ERS classification actually includes 277 nonmetropolitan retirement destination counties as of the 2000 census. However, three of these counties are deleted from our analysis because we only focus on the lower 48 continental states.

6. Figure 2.4 was provided by Ken Johnson. See Johnson (2006) for a discussion of criteria used in the identification of natural decrease counties.

7. This being said, it must be acknowledged that the percentage 65+ in counties that are both RRD and natural decrease was 19.9 percent in 2000 compared with only 19.1 percent for the entire natural decrease category. Accordingly, it is not surprising that these counties experienced natural decrease. This issue will be examined later in this chapter and in Chapter 3.

8. It should be remembered that the percentage 65+ in RRD counties that are also natural decrease counties is 19.9 percent, which exceeds the figure for all natural decrease counties.

9. Their model was a conjoined input–output/econometric model of Wisconsin counties. For a technical discussion of this model see Sheilds and Deller (1997).

REFERENCES

Baker, R., and Speakman, L. (2006). The older rural consumer. In P. Lowe and L. Speakman (Eds.), *The ageing countryside: The growing older population of rural England* (pp. 119–132). London: Age Concern England.

Beale, C. (2005, June 8). Rural America as a retirement option. *Amber Waves*, 8–9.

Brown, D. L., and Kandel, W. (2006). Rural America through a demographic lens. In W. Kandel and D. L. Brown (Eds.), *Population change and rural society* (pp. 3–23). Dordrecht: Springer.

Brown, D. L., and Cromartie, J. (2003). The nature of rurality in postindustrial society. In T. Champion and G. Hugo (Eds.), *New forms of urbanization: Beyond the urban–rural dichotomy* (pp. 269–284). Aldershot, England: Ashgate.

Brown, D. L., and Wardwell, J. (Eds.) (1981). *New directions in urban–rural migration: The population turnaround in rural America.* New York: Academic Press.

Cook, P., and Mizer, K. (1994). *The revised ERS county typology: An overview* (Rural Development Research Report No. 89). Washington, D.C.: U.S. Department of Agriculture, Economic Research Service.

Delanty, G. (2003). *Community.* London: Routledge.

Economic Research Service, U.S. Department of Agriculture. (2004). *Data Sets: County Typology Codes.* Retrieved March 2007, from http://www.ers.usda.gov/Data/TypologyCodes/

Fagin, M., and Longino, C. (1993). Migrating retirees: A source for economic development. *Economic Development Quarterly, 7*(1), 98–106.

Destination Florida Commission. (2003). Destination Florida commission's final report. Retrieved February 2006, from http://www.ccfj.net/DestFlaFinRep.html

Gendell, M. (2006). Full-time work rises among U.S. elderly. Retrieved May 2006, from the *Population Reference Bureau* website http://www.prb.org/Articles/2006/FullTimeWork-AmongElderlyIncreases.aspx

Glasgow, N. (1988). *The nonmetro elderly: Economic and demographic status* (Rural Development Research Report No. 70). Washington, D.C.: U.S. Department of Agriculture, Economic Research Service.

Glasgow, N., and Brown, D.L. (2006). Social integration among older in-migrants in nonmetropolitan retirement destination counties: Establishing new ties. In W. Kandel and D. L. Brown (Eds.), *Population change and rural society* (pp. 177–196). Dordrecht: Springer.

Glasgow, N., and Reeder, R. (1990). Economic and fiscal implications of non-metropolitan retirement migration. *Journal of Applied Gerontology, 9*(4), 433–451.

Golant, S. (2002). Deciding where to live: The emerging residential settlement patterns of retired Americans. *Generations, 26*(2), 66–73.

Haas, W., and Serow, W. (2002). The baby boom, amenity retirement migration, and retirement communities: Will the golden age of retirement continue? *Research on Aging, 24*(1), 150–164.

Haas, W., and Serow, W. (1993). Amenity retirement migration process: A model and preliminary evidence. *The Gerontologist, 33*(2), 212–220.

Halfacre, K. (2003). Rethinking rurality. In T. Champion and G. Hugo (Eds.), *New forms of urban-ization: Beyond the urban–rural dichotomy* (pp. 285–304). Aldershot, England: Ashgate.

Halfacre, K. (1993). Locality and social representation: Space, discourse and alternative representations of the rural. *Journal of Rural Studies, 9*(1), 23–37.

He, W., and Schacter, J. (2003). *Internal migration of older population 1995 to 2000* (Census 2000 Special Reports). Washington, D.C.: U.S. Bureau of the Census.

Johnson, K. (2006). *Demographic trends in rural and small town America.* Carsey Foundation: University of New Hampshire.

Johnson, K. (2005 April). *The rising incidence of natural decrease in rural counties.* Paper presented at the annual meeting of the Population Association of America, Philadelphia, PA.

Johnson, K. (2004 August). *The rising incidence of natural decrease in rural counties.* Paper presented at the annual meeting of the Rural Sociological Society, Sacramento, CA.

Johnson, K., Fuguitt, G. V., Hammer, R., Voss, P., and McNiven, S. (2003 May). *Recent age specific net migration patterns in the United States.* Paper presented at the annual meeting of the Population Association of America, Minneapolis, MN.

Johnson, K., and Cromartie, J. (2006). The rural rebound and its aftermath: Changing demographic dynamics and regional contrasts. In W. Kandel and D. L. Brown (Eds.), *Population change and rural society* (pp. 25–50). Dordrecht: Springer.

Johnson, K., and Rathge, R. (2006). Agriculture dependence and changing population in the Great Plains. In W. Kandel and D. L. Brown (Eds.), *Population change and rural society* (pp. 197–217). Dordrecht: Springer.

Le Mesurier, N. (2006). The contributions of older people to rural community and citizenship. In P. Lowe and L. Speakman (Eds.), *The ageing countryside: The growing older population of rural England* (pp. 133–146). London: Age Concern England.

Longino, C., and Bradley, D. (2003). A first look at retirement migration trends in 2000. *The Gerontologist, 43*(6), 904–907.

McGranahan, D. (2003). How people make a living in rural America. In D. L. Brown and L. Swanson (Eds.), *Challenges for rural America in the 21st century* (pp. 135–151). University Park, PA: Penn State University Press.

Morrill, R., Cromartie, J., and Hart, G. (1999). Metropolitan, urban and rural commuting areas: Toward a better depiction of the United States settlement system. *Urban Geography, 20*(8), 727–748.

Rex, T., and Seidman, L.W. (2002). *Arizona statewide economic study 2002: Retirement migration in Arizona.* Tucson: Arizona Department of Commerce.

Reeder, R. (1998). *Retiree attraction policies for rural America* (Agriculture Information Bulletin No. 741). Washington, D.C.: U.S. Department of Agriculture, Economic Research Service.

Reeder, R., and Glasgow, N. (1990). Nonmetro retirement counties strengths and weaknesses. *Rural Development Perspectives, 6*(2), 12–18.

Ross, P., & Green, B. (1985). *Procedures for developing a policy-oriented classification of nonmetropolitan counties* (Staff Report AGES850308). Washington, D.C.: U.S. Department of Agriculture, Economic Research Service.

Serow, W. (2003). The economic consequences of retiree concentrations: A review of North American studies. *The Gerontologist, 43*(6), 897–903.

Serow, W., and Hass, W. (1992). Measuring the economic impact of retirement migration: The case of western North Carolina. *Journal of Applied Gerontology, 11*(2), 200–215.

Severinghaus, J. (1990). *Economic expansion using retiree income: A workbook for rural Washington communities.* Pullman, WA: Rural Economic Assistance Program, Washington State University.

Shields, M., & Deller, S. (1997 March). *A conjoined input–output/econometric model for Wisconsin counties.* Paper presented to the annual meeting of the Southern Regional Science Association, Memphis, TN.

Skelley, B. D. (2004). Retiree-attraction policies: Challenges for local government in rural regions. *Public Administration and Management, 9*(3), 212–223.

Stallman, J., Deller, S., & Shields, M. (1999). The economic impact of aging retirees on a small rural region. *The Gerontologist, 39*(5), 599–610.

U.S. Office of Management and Budget. (2003). *Revised definitions of metropolitan areas, new definitions of micropolitan statistical areas and combined statistical areas, and guidance on uses of the statistical definitions of these areas* (OMB Bulletin 03–04). Washington, D.C.: U.S. Office of Management and Budget. Retrieved March 2007, from http://www.whitehouse.gov/omb/bulletins/b03-04.html

Vias, A., and Nelson, P. (2006). Changing livelihoods in rural America. In W. Kandel and D. L. Brown (Eds.), *Population change and rural society* (pp. 75–102). Dordrecht: Springer.

Voss, P., McNiven, S., Hammer, R., Johnson, K., and Fuguitt, G. *County-specific net migration by five-year age groups, Hispanic origin, race and sex 1990–2000* (Working Paper No. 2004–24). Madison, WI: Center for Demography and Ecology, University of Wisconsin-Madison.

APPENDIX

List of Rural Retirement Destination Counties as of 1990 Census

Alabama
Baldwin
Cherokee

Arizona
Cochise
Gila
Graham
La Paz
Mohave
Navajo
Baxter
Boone
Carroll
Cleburne
Marion
Sharp
Stone
White

California
Amador
Calaveras
Tuolumne

Colorado
Archuleta
Chaffee
Custer
Delta
Dolores
Fremont
Huerfano
Mineral
Montezuma
Montrose

Ouray

Delaware
Sussex

Florida
Bradford
Calhoun
Citrus
Columbia
DeSoto
Dixie
Flagler
Hamilton
Hardee
Highlands
Holmes
Levy
Madison
Okeechobee
Sumter
Suwannee
Union
Walton

Georgia
Banks
Camden
Candler
Clay
Fannin
Franklin
Gilmer
Greene
Habersham
Hart
Lumpkin

Putnam

Quitman

Rabun

Taliaferro

Towns

Union

White

Hawaii

Hawaii

Idaho

Benewah

Blaine

Bonner

Oneida

Valley

Kansas

Nemaha

Kentucky

Lyon

Menifee

Maine

Lincoln

Maryland

Kent

Worcester

Massachusetts

Dukes

Nantucket

Michigan

Alcona

Antrim

Arenac

Benzie

Charlevoix

Cheboygan

Clare

Crawford

Emmet

Gladwin

Grand Traverse

Iosco

Kalkaska

Lake

Leelanau

Mackinac

Manistee

Mecosta

Montmorency

Ogemaw

Oscoda

Otsego

Roscommon

Minnesota

Aitkin

Cass

Cook

Crow Wing

Douglas

Hubbard

Lake of the Woods

Mille Lacs

Pine

Wadena

Mississippi

Covington

Pearl River

Missouri

Benton

Camden

Hickory

Howell

Iron
Laclede
Miller
Morgan
Stone
Taney
Wayne

Montana
Broadwater
Flathead
Jefferson
Lincoln
Madison
Mineral
Ravalli
Sanders
Stillwater

Nebraska
Hooker

Nevada
Churchill
Douglas
Lyon
Nye

New Hampshire
Belknap
Carroll

New Mexico
Catron
Cibola
Grant
Lincoln
Luna
Mora
Otero
San Miguel

Sierra
Socorro
Taos

North Carolina
Avery
Carteret
Cherokee
Clay
Craven
Dare
Graham
Macon
Moore
Polk
Transylvania

Ohio
Pike

Oklahoma
Delaware
Marshall
McIntosh

Oregon
Coos
Crook
Curry
Douglas
Jefferson
Josephine

Pennsylvania
Forest
Monroe
Sullivan
Wayne

South Carolina
Barnwell

Beaufort
Clarendon
Georgetown
McCormick

South Dakota
Custer
Walworth

Tennessee
Bledsoe
Campbell
Cumberland
Decatur
Fentress
Johnson
Meigs
Monroe
Moore
Putnam
Sevier

Texas
Blanco
Bosque
Burnet
Camp
Coke
Fayette
Franklin
Gillespie
Grimes
Hamilton
Henderson
Hood
Houston
Kerr
Leon
Llano
Madison
Mason

Maverick
Montague
Polk
Rains
Real
San Augustine
Somervell
Starr
Trinity
Van Zandt
Washington
Wood
Young
Zapata

Utah
Daggett
Iron
Kane
Millard
Rich
Sanpete
Sevier
Wasatch
Wayne

Vermont
Bennington
Orleans

Virginia
Accomack
Culpeper
Lancaster
Mecklenburg
Middlesex
Northumberland
Orange
Prince Edward
Richmond

Washington
Clallam
Ferry
Jefferson
Lewis
Mason
Pacific
Pend Oreille
San Juan
Stevens
Wahkiakum

Florence
Forest
Iron
Marinette
Marquette
Oneida
Polk
Vilas
Washburn
Waupaca
Waushara

Wisconsin
Adams
Burnett
Door

Wyoming
Johnson
Lincoln
Teton

CHAPTER 3

THE FORMATION AND DEVELOPMENT OF RURAL RETIREMENT DESTINATIONS

LASZLO J. KULCSAR, BENJAMIN C. BOLENDER AND DAVID L. BROWN

INTRODUCTION

According to the 2000 Census, 277 nonmetropolitan counties had 15 percent or higher population growth at ages 60 and older due to net in-migration during the decade of the 1990s, and hence were classified as rural retirement destinations (RRDs).[1] As will be documented later in this chapter, some of these counties have been continuously categorized as rural retirement destinations since 1979, when the practice of identifying such areas was initiated by USDA's Economic Research Service (ERS), while other counties moved into the category in 1989 or in 2004 when subsequent reclassifications occurred.[2] While considerable research has focused on the demographic and economic impacts of retirement migration, virtually no research has considered why some counties become retirement destinations, why some lose this status, and why some persist in attracting older in-migrants over multiple decades. Accordingly, this chapter focuses on these questions.

Understanding this process is important because retirement destination status has been closely associated with nonmetropolitan population growth over the past three decades. In fact, retirement destinations are one of the only types of rural places to have experienced continuous population growth and net in-migration during this period. It is thought that population growth among rural retirement destinations is associated with certain types of environmental amenities that attract older in-movers, but as we will show in Chapter 4, attractive amenities are only one reason that older migrants give to explain their residential choices. Hence, retirement destinations are not simply places with favorable weather and attractive environmental amenities. Research in this chapter attempts to identify other county attributes that result in relatively high in-migration to rural places by older-aged people.

As the baby boom generation ages, many more retirees will be available to change their places of residence. While we do not know the share of baby boomers that will eventually move—and among these movers what proportion will choose rural destinations (see discussion of this in Chapter 1)—we can be sure that some will indeed move to RRDs. In fact, research has demonstrated that migration among baby boomers is already affecting differential rates of population growth among regions (Plane and Rogerson, 1991) and between metropolitan and nonmetropolitan areas (Nelson, Nicholson and Stege, 2004). Understanding the factors associated with becoming a rural retirement destination will enhance our ability to anticipate and plan for this eventuality.

As indicated in Chapter 2, our research on rural retirement destinations focuses on areas that can be characterized as unplanned retirement destination communities, not places like *Sun City* where an explicit process of planning and development centered around retirement living has occurred. Accordingly, the rural retirement destination counties we study include diverse populations with a range of age groups, economic activities, socio-economic classes and a mix of wealthier and more modest neighborhoods. In this sense, they are more similar to comparable-sized rural communities that lack significant growth in numbers of older residents than they are to planned retirement communities.

The aging process that is occurring in RRDs should be contrasted to what is being experienced by *aging-in-place* rural communities. The latter refers to counties which lack older in-migration but where a large number of persons age 55–64 are surviving into the 65 + age category and where chronic out-migration at younger ages has produced an age structure with low reproductive potential (Fuguitt & Beale, 1993). In contrast to RRDs, aging-in-place communities have both a high and growing concentration of older people because a relatively small proportion of their population is in the child bearing or rearing ages and because they are not attracting in-migrants at either the older or working ages. While the in-migration of older persons is expanding the size of the older population in rural retirement destinations, such areas also typically attract working-age migrants, and they do not *necessarily* have an especially old age structure nor one that is aging rapidly. Of course, it should be recognized that if retirement in-migration persists for a long enough time in RRDs, earlier cohorts of older in-migrants will age-in-place and may produce similar concerns as in those counties where the younger population is not being renewed by either births or in-migration. As will be discussed in this book's concluding chapter, this is seldom recognized or planned for in rural retirement destinations and it may present a significant challenge in the future.

DATA AND METHODS

This chapter uses county-level data from the U.S. Bureau of the Census and a county classification scheme developed by the USDA's Economic Research Service (ERS) to track counties moving into and out of the rural retirement destination (RRD) classification over time. In 1985, the ERS released its classification of counties into what it contended were "policy-oriented classifications" (Ross and Green, 1985). The ERS observed that the classification system reduced the wide range of social and economic diversity in rural America "into a few important themes relevant to rural policy making" (Cook and Mizer, 1994, p. 1). Since the Office of Management and Budget reclassifies counties by metropolitan and nonmetropolitan status after each decennial census, the ERS also updates its nonmetropolitan policy typology consistent with the latest OMB reclassification. While the rules used to identify rural retirement destinations have remained constant over time (15% or higher growth through in-migration at ages 60+),[3] the universe of nonmetropolitan counties from which these classes of counties are identified changes because some nonmetropolitan counties "become" metropolitan and a few metropolitan counties lose their status and are reclassified as nonmetropolitan (Brown and Mauer, 2005).

For this analysis we were able to identify rural retirement destination counties at three points in time corresponding with changes in the USDA-ERS county typology. Even though ERS revised its typology at various times between censuses, we date these three revisions as 1980, 1990, and 2000 because in each instance the revision was based on the decennial census and on the reclassification of metropolitan status based on those censuses. The original classification was based on net migration between 1970–1980 using the 1983 definition of metropolitan and nonmetropolitan status (which in turn was based on the results of the 1980 census). ERS's next revision utilized 1980–1990 net migration data and OMB's 1993 metropolititan–nonmetropolitan reclassification. Finally, ERS's most recent revision was released in 2004 based on the official designation of metropolitan areas that was announced in 2003.[4] This latest revision reflects both migration between 1990 and 2000 and OMB's new core-based system for classifying counties as metropolitan, micropolitan and nonmetropolitan (Brown, Cromartie and Kulcsar, 2004).

We merged county-level social and economic data from the 1970, 1980, 1990 and 2000 decennial censuses along with a county's rural retirement destination status at the three time points.[5] Due to administrative changes over these 30 years, this yielded 3162 counties and county equivalents. We selected only those county units that were constant throughout the whole time period and which were accounted for in the ERS typologies. Hence, this analysis does not include Alaska and Hawaii because the ERS did not develop county typologies for those two states in 1980. This means, for example, that based on the current definition, we lost three of the 277 nonmetropolitan retirement migration destinations identified in 2000. The final number of the counties and county equivalents for this analysis is 3067. The number of nonmetropolitan retirement counties is 480 in 1980, 185 in 1990 and 274 in 2000. Altogether, 584 counties have been classified as rural retirement migration destinations at some point since the first delineation in 1980.

Our first task was to describe possible paths into and out of the changing retirement migration destination category, with special attention given to the impact of metropolitan reclassification. We were particularly interested in whether the counties that lost RRD status did so simply because they "migrated" from the nonmetropolitan to the metropolitan category, as a result of metropolitan reclassification. This is a reasonable expectation because, as indicated above, RRDs as a class have had a relatively high rate of population growth over the last 30 years and rapidly growing rural counties are more likely to become metropolitan. Next we examined the social and economic characteristics of the RRD counties in comparison with the nonmetropolitan total. This comparative analysis was designed to identify county attributes that might be associated with becoming a RRD. We then used logistic regression to determine 1) Why some counties were more likely to be RRDs than others; 2) Why some RRD counties were more likely than others to lose this status; and 3) Why some non-RRD counties were more likely to gain it.

PATHWAYS INTO AND OUT OF THE RURAL RETIREMENT DESTINATION CATEGORY

The Original Cohort's Fate

As indicated earlier, 480 counties were identified as rural retirement destinations in the original 1980 ERS county typology. This number declined to 185 during the next decade, but rebounded to 274 counties in 2000. As shown in Table 3.1, only 111 of the original 480 RRDs maintained this status throughout the 30 years covered by this analysis. These data also show that 245 of the original RRD counties dropped out in 1990 and have not re-entered the category. Another 47 of the original cohort retained their RRD status in 1990 but dropped out during the next decade. In contrast, 77 counties that dropped out in 1990 were reclassified back into the RRD category after the 2000 census. Even though some of them were not in the category continuously, 188 of the 274 RRDs identified in 2000 (69 percent) were retirement destinations at both the beginning and the end of the analysis period. Put somewhat differently, only 86 counties in the 2000 cohort joined the RRD category after the initial classification and almost all of these counties joined between 1990 and 2000. So, even though there has

Table 3.1. Distribution of Counties by Retirement Migration Destination Classification, 1980–2000

Time points			Number of RRD Counties
RRD in 1980[a]	RRD in 1990	RRD in 2000	
Yes	Yes	Yes	111
Yes	No	No	245
Yes	No	Yes	77
Yes	Yes	No	47
No	No	No	2483
No	Yes	Yes	9
No	Yes	No	18
No	No	Yes	77

Source: U.S. Census of Population
[a] RRD County Classification by ERS-USDA.

been considerable movement into and out of this category during the 30 years covered by our study, the majority of current RRDs have been attracting older in-migrants for a considerable time.

The data in Table 3.2 and the map in Figure 3.1 show the location of the original cohort, as well as where counties that dropped out after 1980 are located. As can be seen from this map, almost two thirds of the original 480 counties were concentrated in the South. In fact, Arkansas, Florida, Georgia and Virginia had 20 or more rural retirement destinations each. Seventy-seven were located in Texas alone, situated mainly in the Hill Country. Somewhat surprisingly, almost

Table 3.2. Regional Distribution of RRD Counties, 1980–2000[a]

	1980		1990		2000	
	No.	%	No.	%	No.	%
Midwest	107	22.3	30	16.2	63	23.0
South	277	57.8	95	51.4	125	45.6
West	79	16.5	56	30.3	75	27.4
Northeast	17	3.5	4	2.2	11	4.0
TOTAL	480	100	185	100	274	100

Source: U.S. Census of Population
[a] RRD County Classification by ERS-USDA

Figure 3.1. Nonmetro Retirement Destination Counties 1980, 1990 and 2000

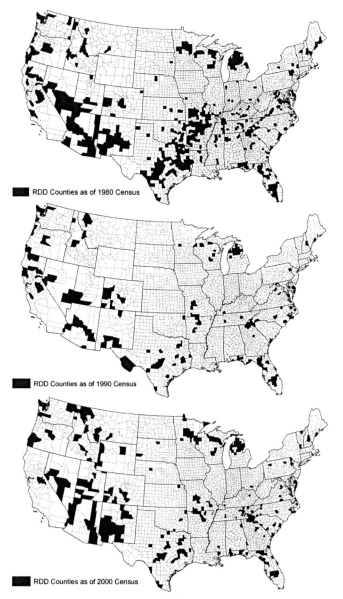

one in five of the original RRDs were located in the Midwest. As indicated on the map, there were conspicuous numbers in Michigan, where outdoor recreation is a big draw, and in Missouri, where the Ozark Mountains are a popular tourist attraction. Only 4 percent of the original group of RRD counties were in the Northeast.

By 1990, the number of nonmetropolitan counties satisfying the 15 percent threshold for migration-induced population growth at ages 60 and older had declined to less than 40 percent of its original number, with only 185 counties satisfying the criterion. Over two thirds of this loss occurred in the South with large losses especially prominent in Texas, Arkansas and Virginia. The Midwest also experienced a significant loss of rural retirement counties, which declined from 107 to 30 during the decade. Michigan and Missouri, the two midwestern leaders in 1980, lost the most counties during this time (see Figure 3.2). While only 17 RRDs were located in the Northeast in 1980, this number fell to 4 in 1990. In contrast to the rest of the country, only 8 percent of the loss in rural retirement counties was located in the West. As a result, the West's share of all RRDs rose from 16.5 percent in 1980 to 30 percent a decade later.

The decade of the 1990s saw a resurgence in the number of rural retirement destination counties, from 185 in 1990 to 274 in 2000 (see Table 3.2 and Figures 3.1 and 3.2). The number of new RRDs increased in all four regions,

Figure 3.2. Counties That Dropped Out of the Original Cohort of RRDs

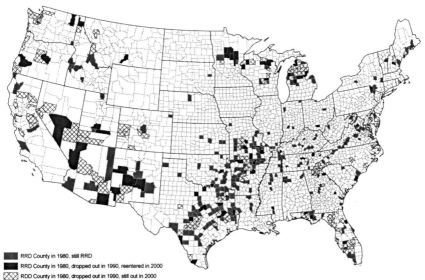

RRD County in 1980, still RRD
RRD County in 1980, dropped out in 1990, reentered in 2000
RDD County in 1980, dropped out in 1990, still out in 2000
Source: ERS-USDA (2004)

but was most notable in the Midwest where it more than doubled during this time. In addition, about one third of new RRDs were in the South (30 counties) and about one fifth in the West (19 counties). The Northeast only added 7 new RRDs, but this was a large increase over its number in 1990. As a result of these new additions, nearly one half all RRDs are now in the South and about one quarter of RRDs are located in both the Midwest and the West. Compared with the location of the original 1980 cohort, RRDs are relatively less concentrated in the South and relatively more concentrated in the West. The Midwest and Northeast have both recovered the shares they had in the initial cohort.

One hundred eleven counties have satisfied the 15 percent growth by migration criterion continuously between 1980 and 2000. As can be seen in Figure 3.3, slightly over half of these counties are in the South, with the other half being roughly evenly distributed in the Midwest and West. Only 3 of the continuous RRD counties are located in the Northeast. Accordingly, while many counties moved into and dropped out of the RRD category between 1980 and 2000, the geographic location of those that persisted in the status over these three decades is quite similar to the distribution of the original group of RRDs. The only difference is that western RRDs showed a somewhat greater propensity to retain their RRD status, compared with their original share of RRDs in 1980,

Figure 3.3. Non-metro Counties in the RRD Category Continuously During 1980, 1990 & 2000

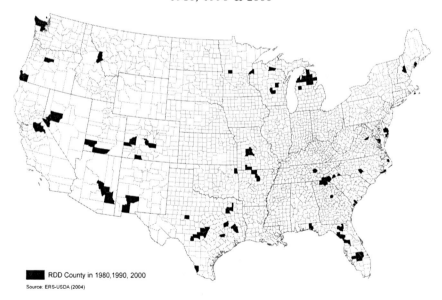

RDD County in 1980,1990, 2000

Source: ERS-USDA (2004)

and southern counties dropped out in slightly higher rates than would have been
predicted by their original dominance.

The Current Cohort of RRDs

The analysis above shows that about 4 in 10 of the current cohort of
RRDs held this status in 1980. However, as shown in Table 3.1, of the 144
"new entrants" to the RRD category between 1990 and 2000, 77 had been in the
category in 1980 while the other 77 were first time RRDs. It is thus somewhat
unusual for "new" RRDs to spring up, or to put it differently, many of the "new"
RRDs in 2000 are re-entering the category rather than being new.

As can be seen in Figure 3.4, current RRDs that did not qualify for
this status a decade before (1990) are mainly located in the upper Midwest and
in the South and Southwest. This shows that the number of RRDs in both of
these regions has rebounded from substantial losses experienced between 1980
and 1990. The South now has 45 percent as many RRD's as it had in 1980,
and this figure rises to 59 percent for the Midwest. This compares with only 34
percent and 28 percent, respectively, when RRDs in 1990 are compared with the
original 1980 cohort. The analysis in Table 3.2 also shows that while the West
experienced a drop off in the number of RRDs between 1980 and 1990, it was

Figure 3.4. Counties Entering RRD Category Since 1980

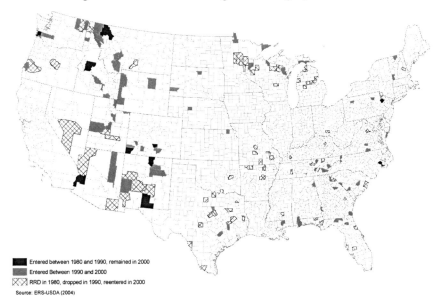

■ Entered between 1980 and 1990, remained in 2000
▓ Entered Between 1990 and 2000
⊠ RRD in 1980, dropped in 1990, reentered in 2000
Source: ERS-USDA (2004)

much less severe than in other regions. Moreover, there are now approximately as many RRDs in the West as there were in 1980. The Northeast has also had a strong rebound, although the number of RRDs in this region is small.

The Effect of Metropolitan Reclassification on RRD Status

As mentioned in the methodology section, the metropolitan status of all US counties is reconsidered after each decennial census. Demographic changes during the inter-census period in combination with any modification of the OMB's metropolitan area criteria have resulted in a substantial number of counties changing from one residential category to the other. During the 1990s, for example, 298 previously nonmetropolitan counties achieved metropolitan status, while 46 previously metropolitan counties moved in the other direction (ERS-USDA, 2004). One result of metropolitan reclassification is that counties which lose their nonmetropolitan status also lose their identification as rural retirement destinations. In other words, if a county is no longer nonmetropolitan, it cannot, by current definition, be a nonmetropolitan retirement destination even if it continues to attract older in-migrants.

Therefore, counties that lose their status as a rural retirement destination can do so by being reclassified into the metropolitan category *or* by having a reduced rate of net in-migration at age 60 and above. Since rural retirement destinations are among the fastest growing nonmetropolitan counties, and since nonmetropolitan counties that experience relatively rapid population growth are more likely to become metropolitan (Johnson and Cromartie, 2006), it is reasonable to expect that metropolitan reclassification might account for a large share of counties that drop out of the retirement destination class between any two censuses. The data in Table 3.3 show that the impact of metropolitan reclassification is lower than anticipated. For example, between 1980 and 1990 only 37 of 322 counties that dropped out of the rural retirement destination status did so because they became metropolitan. The vast majority of dropouts during this time had migration rates at ages 60 and above that failed to satisfy the 15 percent minimum standard.

However, the picture is somewhat different when we focus on the most recent period, 1990–2000. Looking at the 65 counties that were rural retirement destinations in 1990 but lost this status in 2000, slightly over 40 percent of them lost their RRD status because they became metropolitan, not because migration at older ages slowed. Similarly, if we look at the 47 counties that were in the RRD category in both 1980 and 1990 but lost this status in 2000, almost half did so because they were absorbed into the metropolitan category. Accordingly, while metropolitan reclassification was not the major reason for the large drop in the number of RRDs between 1980 and 1990, a substantial

Table 3.3. Impact of Metropolitan Reclassification on Loss of RRD Classification 1980–1990 and 1990–2000[a]

	Lost RRD 1980–1990			Lost RRD 1990–2000		
	Total	Remained Nonmetro	Became Metro	Total	Remained Nonmetro	Became Metro
RRD in 1980[b]	322	285	37	47	25	22
RRD in 1990	NA	NA	NA	65	37	28

Source: U.S. Census of Population
[a] RRD County Classification from ERS-USDA.
[b] Total Number of RRD Counties in 1980 = 480; 1990 = 185.

share of the RRD counties that lost this status in 2000 did so because they were transferred from nonmetropolitan to metropolitan status. While switching out of the nonmetropolitan category may not be as prominent an explanation for the loss of RRD status as we had assumed, it is significant, especially after 1990.

SOCIO-ECONOMIC AND DEMOGRAPHIC CHARACTERISTICS OF RRD COUNTIES IN COMPARISON WITH OTHER RURAL AREAS

As indicated earlier, RRD counties have grown more rapidly than the rest of rural America. The data in Table 3.4 show that the current group of RRD's grew by 13.8 percent between 1980 and 1990 compared to a loss of 2.7 percent for other nonmetropolitan counties during that time. Between 1990 and 2000, RRDs continued to have a substantial growth advantage. They grew by 25.2 percent compared with 4.7 percent for other nonmetropolitan counties. The data in Table 3.4 show that higher migration rates among RRDs contribute to this growth differential. As a partial result of higher rates of growth and migration since 1970, RRDs which were smaller than other nonmetropolitan counties in 1970 are now substantially larger.[6] It is interesting that this is true regardless of whether the comparison of migration and population growth rates is done with the original cohort of RRDs or with the most recent cohort.

In contrast to the substantial differences in population growth and migration discussed above, RRDs are not much different from other nonmetropolitan counties with respect to demographic composition and socio-economic status. Regardless of which indicator is viewed, differences are modest. As indicated previously, RRDs attract older in-migrants, but they also attract younger adults and their families. Accordingly, while RRDs are slightly older

Table 3.4. Comparative Profile of RRD and Non-RRD Nonmetropolitan Counties, 1980–2000

	As of 1980 Census		As of 1990 Census		As of 2000 Census	
	RRD	Non-RRD	RRD	Non-RRD	RRD	Non-RRD
Average population, 1970	18,022	20,577	15,446	19,519	14,594	20,022
Average population, 2000	34,249	24,443	33,897	23,616	27,962	23,311
Percent population change 1980–1990	13.1	−1.9	22.1	−1.6	13.8	−2.7
Percent population change 1990–2000	21.4	6.3	28.5	7.0	25.2	4.7
Percent lived in differ. county 1995	21.6	17.5	24.4	17.7	23.0	16.0
Percent 65+, 2000	16.6	15.4	17.9	15.6	17.2	15.9
Percent nonwhite, 2000	16.7	17.5	17.1	17.2	16.1	17.3
Percent some college, 2000	40.7	40.0	45.2	39.5	43.0	39.4
Percent unemployed, 2000	3.3	3.3	3.3	3.4	3.5	3.4
Median household income, 1999	33,296	32,454	33,897	32,118	33,226	31,647
Percent poverty, 1999	14.1	14.8	13.3	14.9	13.8	15.2
Natural amenity index	1.3	−.35	2.6	−.3	1.7	−.3
Percent South & West	74.2	55.6	81.7	56.9	73.0	54.6
Percent adjacent to SMSA[a]	48.5	42.2	52.4	43.6	61.7	50.5
Percent college located in county, 1990	29.8	31.3	27.0	30.4	26.6	31.8

Source: U.S. Census of Population, Amenity Index McGranahan (1999)
[a] SMSA as defined after respective censuses

than other nonmetropolitan counties, differences in age composition are not large. Similarly, about 8 out of 10 persons in nonmetropolitan counties are white, regardless of RRD status. Since we know that 90 percent of America's rural Black population is concentrated in the South, and that about half of current RRDs are also located in this region, it is clear that RRDs are not randomly spread throughout this region with respect to race. Rather, they are concentrated in predominately white areas. Retirement migration appears to avoid the

Mississippi Delta and the Black Belt, southern regions with a historic legacy of concentrated Black settlement.

RRDs have slightly higher rates of college educational attainment[7] and higher median household income, as well as slightly lower poverty rates. However, these differences, while systematic, are not large. As with other issues, these comparisons of socio-economic status are *not* sensitive to whether we compare the original cohort of RRDs with the rest of rural America or use the later cohorts for the comparison.

Finally, we compare RRDs with other nonmetropolitan counties with respect to several indicators of the county as a place. As can be seen in Table 3.4, RRDs are higher in natural amenities than other nonmetropolitan counties.[8] This is consistent with the general observation that natural amenities are one of the attributes that attract older in-migrants to RRDs. Finally, RRDs are significantly more likely to be located next to a metropolitan county and this pattern has grown stronger over time. Only 49 percent of the original cohort of RRDs were contiguous to a metropolitan county as officially classified at that time. This figure has risen to 62 percent in 2000. In contrast, only half of other nonmetropolitan counties are currently adjacent. As will be shown in the next chapter, many of the older in-migrants to rural retirement destinations we interviewed cited access to amenities, facilities, services and nearby relatives as important considerations motivating their choice of new communities, and many of these are located in the neighboring SMSA. One surprising comparison in Table 3.4 is that RRD counties are somewhat *less* likely to be the location of an institution of higher learning than is true of other nonmetro counties but, again, this difference is small (27 percent vs. 32 percent). Many of the persons we interviewed in rural retirement destinations indicated that the presence of a college was attractive to them, but as shown in Table 3.4, presence of a college is not a community attribute that is likely to differentiate retirement destinations from other rural places. This lack of effect is also supported by the multivariate statistical analysis that will be presented later in this chapter.[9] Regional location is the largest difference between RRD and non-RRD counties shown in Table 3.4. Three quarters of RRDs are in the South and West compared with only about half of other nonmetro counties.

FACTORS ASSOCIATED WITH GAINING, KEEPING AND LOSING RRD STATUS

Rural retirement destinations do not simply sprout from the ground in a random fashion. Some types of places are more likely than others to attract older persons and assume this status. This section explores why some nonmetropolitan counties are more likely than others to become or persist as retirement destinations, and why some retirement destination counties are more likely than others to lose this status. While interviews with older migrants themselves offer insights into the place attributes that attracted them to particular rural counties, these individual-level migrant accounts do not provide a systematic explanation of the social, demographic, economic and amenity attributes that advantage some counties over others with respect to attracting older persons. Many observers have commented that RRDs are environmentally attractive, have warm weather, relaxed life styles and a high "quality of life". However, no previous research has used a multivariate approach to systematically examine the impact of these and other factors in predicting why some counties are more likely than others to be RRDs, or to gain or lose this status. Accordingly, this section attempts to develop a holistic explanation of the demographic, socio-economic, and environmental factors which are associated with the chances that a rural county will become and stay a retirement destination. Our model includes four domains of county-level variables which are hypothesized to increase or decrease the odds that a county will be characterized as a rural retirement destination, or that it will gain or lose this status.

The independent variables were categorized into four sets. The first three reflect a county's socio-demographic and economic characteristics. The fourth block reflects place characteristics including environmental attributes, geographic situation and major economic activities. The four blocks of variables are: (1) Population size, change and urbanization; (2) demographic composition; (3) Socio-economic status; and (4) Geographic context and principal economic sectors. These blocks of variables are added sequentially, and then considered together in a full model. Most variables have both static and dynamic versions, where the static refers to an observed characteristic at the previous census and the dynamic is the change in that variable comparing a census year to the one that preceded it. Thus, dynamic variables for the 2000 typology refer to change between 1990 and 2000. The natural amenity scale was kept consistent for all three time periods, assuming that topographic features remain constant, and climate is relatively stable over the short time period considered in this study. The college variable was also kept consistent because these data were only available at one point in time. As discussed in the methodology section, the adjacency-non-adjacency to Standard Metropolitan Statistical Area (SMSA) variable and the county typologies were revised by ERS in 1990 and 2000.

Factors Associated with Being a RRD in 2000

We use logistic regression to examine factors associated with the likelihood that a nonmetropolitan county will be a RRD in 2000. The outcome variable for this analysis is: *RRD counties in 2000 versus other nonmetropolitan counties that are not RRDs in 2000*. Because being a RRD involves gaining older in-migrants during the decade prior to obtaining, or retaining, the official status, predictor variables represent the *beginning* of the decade (1990) or they measure changes that occurred between the 1990 and 2000 censuses. Our investigation involves the four domains of factors discussed above. In addition, since we know that some current RRDs were in the category in 1980, but not in 1990, we include a control for whether a county has ever been a RRD in the past. The data in Table 3.5 show how these factors affect the chances that a rural county will be a RRD.[10] Blocks associated with the four domains of variables are added one after another, followed by the control for whether a county was ever a RRD prior to 2000. The odds ratios presented in Table 3.5 show whether an independent variable increases or decreases the chances that a nonmetropolitan county will have RRD status in 2000. A ratio greater than 1.0 indicates a positive impact, while an odds ratio less than 1.0 shows that the factor reduces the chances.

As shown in the first column of Table 3.5, demographic size, change and urbanization variables have a measurable impact on whether a county is a RRD in 2000. To begin with, more highly urbanized nonmetropolitan counties are *less* likely to be RRDs than their less urbanized counterparts. On the other hand, counties that grew in population and those that gained in-migrants from elsewhere in the same state were more likely to be RRDs. However, migration gains from other states and/or from abroad, while in the expected positive direction, had no statistically significant impact on the odds that a county would be a RRD. These three factors explain about 40 percent of the likelihood of being a RRD in 2000.

We next entered the block of variables indicating a county's age and racial composition. Our expectation was that older in-migrants would tend to move to areas that had a demographic composition most reflective of their own characteristics, such as places with relatively high percent 65+ and that were primarily white. The results in Table 3.5 indicate that our expectations regarding age were reasonable. Counties with a relatively high percent of their population aged 65 and above are more likely to be RRDs. Moreover, counties that increased in percent 65 are over three times more likely to be RRDs. In contrast, the effects of race are not consistent with our expectations. Neither percent nonwhite nor rate of growth in nonwhite population was statistically associated with RRD status. It should also be noted that adding socio-economic factors to the model erases the negative effect of urbanization shown in model 1; and this factor remains insignificant throughout the remainder of the analysis. It is likely that urbanization and racial composition share a slight negative effect on being a RRD, but when they are considered together neither is a statistically significant predictor of RRD status. Adding age and racial composition to the model also

Table 3.5. Factors Associated with RRD Status in 2000

	Odds Ratios					
	Model 1	Model 2	Model 3	Model 4	Model 5	Model 6
Population size, change and urbanization						
Population 1990[a]	1.008	1.006	.997	.999	.992	.991
Migration same state[b]	1.042*	1.049*	1.011	1.050	1.038	1.033
Migration different state[c]	1.014	.988	.980	.998	1.003	.997
Percent urban	.978***	1.004	1.010	1.007	1.006	1.006
Percent population change, '90-'00	1.123***	1.247***	1.260***	1.251***	1.222***	1.216***
Change in Percent urban, '90-'00	.991	1.001	1.002	1.000	1.003	1.003
Demographic composition						
Percent 65+		1.529***	1.470***	1.496***	1.400***	1.343***
Percent nonwhite		1.012	1.003	1.008	1.010	1.010
Change in Percent 65+, '90-'00		3.033***	3.100***	3.202***	2.992***	2.918***
Change in Percent nonwhite, '90-'00		.987	.971*	.982	.979	.974
Socio-economic status						
Percent with BA degree or higher			.995	.958	.961	.949
Percent employed			.880***	.901***	.906***	.916***
Percent poverty			.975	1.003	.995	.990
Change in Percent college degree			.998	.968	.972	.963
Change in Percent employed			.991	.984	.957	.967
Change in Percent poverty			1.007	1.050	1.032	1.012
College located in county			1.567	1.514	1.580	1.703

Geographic context and principal economic sectors						
Adjacent to SMSA[d]				.896	.978	.977
Farming dependent county[e]				.259***	.332**	.355***
Government dependent county				.780	.795	.827
Recreation dependent county				1.945*	1.751*	1.761*
Natural amenity index				.933	.899	–
South				.783	.640	.647
Previous status as RRD						
Previous status					3.725***	3.552***
−2 log likelihood	1064.806	742.082	696.694	673.147	642.736	645.710
Nagelkerke R²	.426	.633	.660	.673	.691	.689

Source: U.S. Census of Population; *p < 0.05; **p < 0.01; ***p < 0.001.
[a] All variables are for 1990 unless otherwise stated; [b] Migration is for 1995–2000;
[c] Includes from abroad; [d] Adjacent to SMSA as of 1990 OMB classification;
[e] USDA-ERS economic dependency statuses are as of the 2000 classification cycle (farming, government, & recreation).

slightly increases the strength of the migration variable. The importance of adding age composition to the model is indicated by the summary statistics. The Nagelkerke R^2 has now increased from .43 to .63 and the -2 log likelihood statistic declined substantially, showing that age composition and change therein are important predictors of the likelihood that a nonmetropolitan county is a RRD.

Socio-economic status variables were added in block 3. Our general expectation was that older in-movers would be attracted to better off counties. The data do not confirm this expectation. Older in-migrants do not seem to be attracted to places with higher than average rates of college education nor do older in-migrants seem to avoid counties with higher than average poverty rates. In contrast, counties are less likely to be RRDs if they have higher employment rates. While this may seem counter-intuitive, it should be remembered that adding retired persons to a county's population adds persons to the denominator of the employment rate, but many fewer to the numerator.

Accordingly, this deflates employment rates in places with a relatively high in-movement of older persons, most of whom have exited the work force. This block of variables also includes whether the county is the location of at least one institution of higher education. The case studies, described in Chapter 6, that we conducted in four rural retirement destinations indicate that older in-movers are attracted by the cultural and athletic programs offered by colleges and other institutions of higher learning. Moreover, there is anecdotal evidence that some older persons feel a strong connection with their alma mater and choose to retire where they went to college. However, while the presence of a college is in the expected positive direction, it is not statistically significant. This is probably because the proportion of RRD and non-RRD counties with a college does not differ greatly (Table 3.4). It should be noted that the addition of the socio-economic status variables erases the positive impact of intra-state migration shown in models 1 and 2. Accordingly, only age composition and population change remain statistically significant and substantively important predictors of RRD status at this point in the analysis.

Block 4 contains aspects of the county's geographic situation, economic structure and natural environment. Counties whose economies are heavily dependent on recreation and tourism are almost twice as likely to attract older in-migrants compared with counties which lack this economic base. Clearly, older in-movers enjoy the opportunities for leisure activities offered in such places. In contrast, older in-movers are not likely to be attracted to counties that are heavily dependent on agriculture. Such places are mainly in the Great Plains, offer few environmental or cultural amenities and lack access to nearby larger places. Moreover, agriculturally-dependent places are experiencing overall population decline, which was shown earlier in this analysis to negatively effect the odds that a county would be a retirement destination.

Counties with a high rating on the natural amenities scale are no more likely to be a RRD than their counterparts with lower scores. This result may seem counterintuitive because so much retirement development occurs in the "sun belt" and in areas with hills, lakes and shoreline. However, it should be remembered that our study focuses on unplanned retirement communities, not on *Sun City* or *Leisure World*, so the selectivity of warm winter weather and well manicured landscapes is not so strong. While we are not certain why natural amenities fail to predict RRD status, we would observe that favorable amenities to one person may be unattractive to others. Some people prefer warm winters while others like abundant snow. Some people like hills and mountains while others feel more comfortable in flat environments. Since the amenities variable used in this analysis is a composite scale, it is possible that certain of the scale's component attributes counterbalance each other.

One might assume that the non-significance of the amenities variable is because counties with high amenity ratings are also likely to be recreation and tourism areas. Many researchers make this assumption, but the two variables only correlate at $r = .26$ across U.S. nonmetro counties. While the natural amenity scale is related to climate and topography (McGranahan, 1999), the recreation county classification is derived from the amount of employment and income generated by recreation and tourism and the percent of housing used for seasonal or recreational purposes (Johnson and Beale, 2002).[11] Accordingly, it is possible that only certain types of natural amenities tend to attract retirement-age migrants or that some types of natural amenities are more likely to be developed for recreation and tourism than others. Retirement-oriented migration networks form in areas where people vacation or own second homes regardless of whether these areas have warm weather, low humidity, diverse topography, abundant coastline or other natural attributes. These networks then increase the chances for retirement in-migration to be sustained in future years. We wondered if deleting the natural amenities variable from the analysis would affect the model. As shown in the last column in Table 3.5 (model 6), deleting natural amenities has virtually no effect on the results reported when amenities are included (model 5). So, while the natural amenities variable does not produce easily interpretable results, deleting it from the model does not change the effects of other factors.

Being located in the South does not affect the odds that a county is a RRD.[12] This is surprising because 46 percent of RRDs are located in this region. The regional effect is probably diffused by factors like recreation that are concentrated in the South. We were also surprised that counties which are adjacent to a SMSA were no more likely to be RRDs than non-adjacent counties. We thought that adjacent counties would be especially attractive to older in-movers because of the ease of access to cultural amenities, advanced medical services and adult children who may live in the nearby city, but the data in

Table 3.5 do not substantiate this expectation. In fact, while not statistically significant, the effect of adjacency is actually negative.

Finally, we added a control variable to account for the fact that the 2000 RRD category is diverse with respect to whether counties have been in the RRD category continuously since 1980, were ever RRDs in the past, or are new entrants for the first time in 2000. The data in Table 3.5 show that being a RRD at any time previous to 2000 raises the odds of being a RRD in 2000 by over three times! Regardless of whether RRD status has been continuous, counties with previous membership in the category are much more likely to be RRDs in 2000. This is especially impressive since all of the other predictors of being a RRD have already been accounted for in the model. It appears that once a rural county begins attracting older migrants at a rate sufficient to qualify as a retirement destination, social networks are developed. These networks link the area with potential migrants, which results in a long term, self-perpetuating trend of older in-migration. Hence, even if the rate of older in-migration diminishes at some time after a county first qualifies as a RRD, it is likely to increase again once period- or place-specific conditions which might diminish in-migration have subsided.

Many of our expectations were not confirmed by the multivariate analysis, but we feel that our model explains some of the reasons why some nonmetropolitan counties are more likely to be RRDs than others in 2000. Even though the Nagelkerke R^2 is not the same as a R^2 statistic in ordinary regression analysis, it gives an indication of how well the model fits the data. After the first block of variables was added the Nagelkerke R^2 was .426, and it rose to .691 in the full model. Moreover, the -2 log likelihood statistic declined from 1064.806 after the first block was entered to 642.736 in the full model, which shows that adding subsequent variables improves the model's explanatory power. To summarize our results, compared with nonmetro counties that are not RRDs in 2000, rural counties that have at least 15 percent population growth from in-migration at ages 60+ grew more rapidly during the 1990–2000 decade, were older in 1990, aged more rapidly during the 1990s, had a lower employment rate, were more likely to have been a RRD sometime in the past and had a higher level of employment dependence on recreation and tourism and a lower level of dependence on agriculture. This county-level analysis is consistent with many of the responses older migrants offer when asked why they selected their new residences in rural retirement destinations; and it illuminates other issues not expressed by in-migrants when they are asked to explain their residential choices.

Perhaps the most important finding relates to age composition. It appears that once older persons begin moving to a county, the process becomes self-sustaining. Counties with increasing percents aged 65 and above are about three times more likely than counties that are not aging as rapidly to attract retirees. This is an especially impressive finding because some of the negative effects of aging-in-place are controlled in the model through the inclusion of the farming dependence measure. While rural retirement destinations and recreation dependent areas are not the same, it is clear from this analysis that the recreation and retirement are closely linked. Counties with high dependence on recreation and tourism employment are almost twice as likely to attract older in-migrants as counties with lower dependence on these industries.

Why Did Some 1990 RRD Counties Drop Out Between 1990 and 2000?

The above analysis shows why some counties are more likely than others to be RRDs in 2000, but it does not explain why counties enter and exit the category. As shown earlier (Table 3.1), only 120 of 185 counties that were RRDs in 1990 remained in the category in 2000. Of the 65 drop outs, 20 dropped out because they became metropolitan, but they continued to have 15 percent or higher rates of population growth from in-migration at age 60+.[13] Even though these 20 counties are no longer officially nonmetropolitan, we retain them in the RRD category in this analysis because they are still attracting older in-migrants at a higher than average rate. In other words, they are still retirement destinations; and, even if they are now officially metropolitan, they are mostly low density peripheral additions to existing SMSAs. Accordingly, it is not inconsistent with the RRD concept to still consider them to be rural retirement destinations, as we indeed do for our next analysis. The outcome variable for this analysis is *whether a RRD county in 1990 dropped out of the category between 1990 and 2000.* We regressed this variable on the same four blocks of predictor variables that we examined in the preceding analysis of RRD status in 2000. We argue that the forces that explain why some counties are in the RRD category at a particular time should also explain why some RRD counties exit the category, except that the signs of the coefficients would be reversed. The results of this analysis are presented in Table 3.6. Note that in this analysis an odds ratio greater than 1.0 indicates that counties are *more likely to drop out* of the RRD category, while a ratio less than 1.0 indicates that a county is *less likely to drop out.*

The data in Table 3.6 show that growing and more highly urbanized RRDs are less likely to drop out of the category compared with places that are demographically stagnant. An older age composition also reduces the chances of dropping out. The positive effect of population growth is consistent with the findings presented earlier, but the protective impact of urbanization is not consistent with the earlier analysis. Evidently, retirement age migration to more urbanized places is more likely to be maintained over time, perhaps because such places have a wider complement of health and other types of services needed

Table 3.6. Factors Associated with Losing RRD Status between 1990 and 2000

	Odds Ratios				
	Model 1	Model 2	Model 3	Model 4	Model 5
Population size, change and urbanization					
Population 1990[a]	.981	.998	1.018	1.017	1.015
Migration same state[b]	.947	1.019	1.091	1.115	1.175
Migration different state[c]	.982	1.058	1.108*	1.143*	1.168*
Percent urban	1.005	.953*	.944*	.919**	.940*
Percent population change, '90-'00	.913***	.798***	.756***	.642***	.700***
Change in percent urban, '90-'00	1.003	.974	.954	.896*	.937*
Demographic composition					
Percent 65+		.626***	.688**	.471**	.514**
Percent nonwhite		1.028	1.023	1.040	1.011
Change in percent 65+, '90-'00		.313***	.271***	.125***	0.174***
Change in percent nonwhite, '90-'00		1.071*	1.083	1.155*	1.121
Socio-economic status					
Percent with BA degree or higher			1.153	1.059	1.190
Percent employed			1.268*	1.443*	1.215
Percent poverty			1.285*	1.199	1.284

	Model 1	Model 2	Model 3	Model 4	Model 5
Change in percent college degree			.766*	.598**	.745*
Change in percent employed			.972	1.153	.932
Change in percent poverty			1.128	1.086	1.170
College located in county			.126*	.186	.135*
Geographic context and principal economic sectors					
Adjacent to SMSA[d]				.533	1.324
Farming dependent county[e]				.409	.638
Government dependent county				.104	.090
Recreation dependent county				6.424	4.037
Natural amenity index				2.104**	
South				20.461*	4.469
−2 log likelihood	166.842	108.733	87.328	66.109	79.012
Nagelkerke R^2	.280	.607	.703	.789	.738

Source: U.S. Census of Population; * $p < 0.05$; ** $p < 0.01$; *** $p < 0.001$.
[a] All variables are for 1990 unless otherwise stated; [b] Migration is for 1995–2000;
[c] Includes from abroad; [d] Adjacent to SMSA as of 1990 OMB classification;
[e] USDA-ERS economic dependency statuses are as of the 2000 classification cycle (farming, government, & recreation).

to sustain retirement-oriented development. Counties with older populations and those that are aging more rapidly are less likely to drop out of the RRD category. Older in-migrants seem to seek out places with a high and growing density of older persons. As we observed in the analysis of being a RRD in 2000 (Table 3.5), once older persons begin to move to a place, migration appears to be self-sustaining. RRD counties that have an increased representation of persons with a college degree are also less likely to exit the RRD category. This is probably because such places are gaining skilled workers who provide essential services to the growing older population. This is consistent with the positive impact of population growth and increased levels of urbanization reported above. In other words, older in-migrants seem to continue to prefer counties that are experiencing an overall process of development.

While percent nonwhite, in and of itself, has no impact on the likelihood of dropping out of the RRD category, counties that are becoming increasingly racially diverse are somewhat more likely to leave the RRD category. It is possible that retirees, who are mostly white, avoid areas with growing nonwhite populations, regardless of those areas' initial levels of racial diversity. Similarly, higher than average in-migration from other states and from abroad is also positively associated with dropping out of the RRD category. Interstate and international migration may increase racial and ethnic diversity which also increases the likelihood that a county will lose its RRD status.

Surprisingly, RRD counties that have higher than average scores on the natural amenities scale are more likely to lose their RRD status. We are at a loss to explain this result. We can think of no reason why counties with desirable weather, diverse topography and abundant water resources should stop attracting older in-migrants. Next, we added location in South as a control because we know that 24 out of the 45 drop outs are in this region. Predictably, even after demographic, socio-economic and contextual factors are accounted for, southern RRDs are much more likely to drop out of the category.

Similar to the analysis of current RRD status reported in Table 3.5, we ran the RRD exit model without the natural amenities variable. As shown in model 5 (the last column of Table 3.6), in this instance deleting the natural amenities variable does alter the results. First, running the model without the amenities scale increases the negative impact of being a college town on the likelihood of dropping out of the RRD category. In other words, once excluding the amenities scale, we find that the presence of a college contributes to being able to maintain a strong positive flow of older in-migrants. Second, deleting the amenities variable substantially reduces the size of the South effect and diminishes its statistical impact below the level of significance.

Our analysis provides insights into factors affecting the loss of RRD status between 1990 and 2000 among counties that were in the category in

1990. Indicators of overall development including population growth, increased urbanization, an increase in the college educated population, the presence of a college and rapid aging all reduce the chances that a RRD will exit the category. In contrast, increased migration from out of State and a growing nonwhite population increase the chances of dropping out. However, several of the factors shown to be important predictors of current status as a RRD are not significantly associated with an increased likelihood of exiting the category after 1990. Dependence on recreation and tourism employment, for example, predicted RRD membership; but a low score on this variable is not associated with a greater likelihood of dropping out of the category during the decade of the 1990s. In addition, dependence on agriculture does not increase the chances of dropping out as might be expected, but this is likely because few if any RRD counties in 1990 were farming-dependent to begin with.

Why Did New Counties Enter the RRD Category Between 1990 and 2000?

While 65 counties dropped out of the RRD category between 1990 and 2000, their loss was more than compensated for by the 154 new entrants during this period.[14] In this analysis we predict the *likelihood that a rural county that was not a RRD in 1990 became one in 2000*. The results are shown in Table 3.7. Positive odds ratios indicate that the likelihood of becoming a RRD is enhanced by the particular factor, while odds ratios less than 1.0 indicate the opposite. The results of this analysis are very similar to those which explained why some nonmetro counties are more likely than others to be RRDs in 2000 (see Table 3.5). The data in Table 3.7 show that relatively rapid population growth, population aging and dependence on recreation and tourism are the only factors that increase a county's chances of becoming a new RRD between 1990 and 2000. Conversely, it is only dependence on agriculture that reduces the chances of joining the category. The fact that a high initial and increasing percentage of persons aged 65+ is associated with gaining RRD status during the 1990s is not entirely surprising, as there is a built-in association between these factors. Even though RRDs are not exceptionally old, in-migration at ages 60 and older does marginally increase the percent at these advanced ages. While we showed in Table 3.6 that being a college town seems to help RRD counties retain their status, having a college or other institution of higher learning does not contribute to a county's chances of entering the category. Counties with higher employment rates and increasing employment rates are less likely than other nonmetropolitan counties to become RRDs, but as explained earlier this is a statistical artifact. Attracting older persons adds people to the denominator of the employment rate, but because older in-migrants are largely retired, not many are added to the numerator. Accordingly, a negative association is built into the relationship

Table 3.7. Factors Associated with Gaining RRD Status Between 1990 and 2000

	Odds Ratios				
	Model 1	Model 2	Model 3	Model 4	Model 5
Population size, change and urbanization					
Population 1990[a]	.997	1.005	.996	.999	.999
Migration same state[b]	.994	1.022	.984	1.027	1.030
Migration different state[c]	1.008	1.004	.992	1.006	1.010
Percent urban	.986**	1.000	1.005	1.002	1.002
Percent population change, '90-'00	1.058***	1.140*	1.166***	1.143***	1.146***
Change in percent urban, '90-'00	.995	1.000	.999	.999	.999
Demographic composition					
Percent 65+		1.407***	1.385***	1.399***	1.401***
Percent nonwhite		1.016*	.994	.998	.998
Change in percent 65+		2.225***	2.358***	2.279***	2.314***
Change in percent nonwhite		1.004	.974	.981	.984
Socio-economic status					
Percent with BA degree or higher			1.038	.987	.992
Percent employed			.849***	.882**	.879***
Percent poverty			.981	1.010	1.011

	M1	M2	M3	M4	M5
Change in percent college degree			1.035	.978	.982
Change in percent employed			.881**	.884***	.881***
Change in percent poverty			.981	1.005	1.010
College located in county			1.675	1.683	1.628
Geographic context and principal economic sectors					
Adjacent to SMSA[d]				.799	.742
Farming dependent county[e]				.305**	.293**
Government dependent county				.713	.708
Recreation dependent county				3.337***	3.307***
Natural amenity index				1.052	
South				.931	.931
−2 log likelihood	927.931	785.041	711.940	677.050	677.694
Nagelkerke R^2	.150	.307	.383	.419	.418

Source: U.S. Census of Population; * $p < 0.05$; ** $p < 0.01$; *** $p < 0.001$.
[a] All variables are for 1990 unless otherwise stated; [b] Migration is for 1995–2000;
[c] Includes from abroad; [d] Adjacent to SMSA as of 1990 OMB classification;
[e] USDA-ERS economic dependency statuses are as of the 2000 classification cycle (farming, government, & recreation).

between older in-migration and changes in the percent of the population that is employed. While counties that are highly dependent on recreation and tourism employment are over three times more likely to gain RRD status, farming dependent counties are very unlikely to become a RRD. Finally, similar to the results shown in Table 3.5, while one third of new entrants during 1990–2000 were in the South, being in the southern region has no statistical impact on the chances that a county joined the category during the 1990s; nor does having a high score on the natural amenities scale. We deleted the amenities scale from the analysis once again (model 5), and similar to what was shown in Table 3.5, this deletion does not affect any of the other results presented in model 4.

The analysis displayed in Table 3.7 is a first step toward examining why some rural counties become retirement destinations. However, when all of the factors we have examined are accounted for, we are only able to observe that growing counties, especially those with larger and increasing shares of older persons, and places that have well developed recreation and tourism industries are more likely to enter the RRD category. In contrast, farming dependent counties are highly unlikely to become a retirement destination. The Nagelkerke R^2 for this model is only .418 showing a weak to moderate fit with the data.

Retirement Migration, Aging-in-Place, and Natural Decrease

As was shown in our multivariate analysis, percent 65+ and growth in percent 65+ are the most consistent predictors of being or becoming a RRD and of retaining RRD status once in the category. This raises the question of causal direction between increased population aging and becoming a RRD. We believe the relationship works in both directions. Once established, migration streams become self-sustaining. Older migrants are likely to move to places where retirement is an established way of life. Hence, having a relatively high and increasing share of older persons attracts additional older persons to a community. In contrast, continuous in-movement of older persons to rural retirement destinations also elevates average age because many in-movers age in place in their new communities. Accordingly, aging-in-place of previous older in-migrants will eventually increase the possibility of natural decrease even if older in-migration also stimulates in-movement at younger ages.

The Overlap of Retirement Migration and Natural Decrease

As shown in Chapter 2, there is a considerable overlap between RRDs and natural decrease counties (see Figure 2.5). While only 17 percent of nonmetro natural decrease counties are rural retirement destinations (141 of 837), over half of RRDs experienced natural decrease between 1990 and 2000 (141 of

274). This coincidence is surprising since the percent 65 and older among RRDs is only slightly higher than that in non-RRDs (17.2 percent vs. 15.9 percent). However, if one compares counties that have held RRD status continuously since 1980 with those that entered or re-entered the category during 1980–2000, one gains insights into the developmental process that results in counties entering the RRD category. As shown in the previous chapter (see Table 2.1), 68 of the 111 continuous RRD counties experienced natural decrease during 1990–2000. These 68 counties are considerably older than the average RRD (19.9 percent 65 and above versus 17.2 percent 65 and above). In fact, they are even older than the overall natural decrease category (19.9 percent 65 and above versus 19.1 percent 65 and above). While not direct evidence, these data strongly suggest that a large share of older persons who move to rural retirement destinations age-in-place in these locations.

The overlap of RRD and natural decrease counties was surprising to us at first because our image of natural decrease counties was that they were almost exclusively located in heavily agricultural areas in the Northern Great Plains and Corn Belt. We thought that natural decrease areas and rural retirement destinations were essentially different social and economic worlds. However, not all natural decrease counties are located in the Great Plains; and, irrespective of geography, retirement migration that persists over a long enough time can result in natural decrease.

The fact that many retirement destinations are experiencing natural decrease is not on the current policy agenda. This is partly because developmental theories of elderly migration predict that many older in-migrants will be temporary residents who move on once they reach advanced old age and/or experience an adverse life course transition, such as losing their spouse or becoming disabled. As proposed by Litwak and Longino (1987), the developmental theory of elderly migration predicts that many older amenity-seeking migrants will leave retirement destinations later in life to be closer to their adult children. Our data are not consistent with this theory. As will be shown in Chapter 5, over one third of respondents to our panel survey of older in-migrants to rural retirement destinations had at least one adult child within a one half hour drive. Both amenity and family factors seem to be motivating their migration to rural retirement destinations. It appears that many older in-migrants to RRDs are there to stay.

CONCLUSIONS

This chapter examines why some rural counties are more likely than others to be retirement destinations. Understanding this process is important because retirement destination counties are one of the only type of rural areas to experience continuous population growth during the last 35 years. Moreover, as the large baby boom cohorts approach retirement age, a portion of these retirees will undoubtedly change their residences and some will move to rural areas. Hence, attracting older in-migrants can be an integral aspect of more general growth and development. In fact, as indicated in Chapter 2, many states have explicit strategies to recruit older persons as an impetus to rural economic development. Understanding the social, economic and demographic dynamics underlying this process contributes to understanding the ramifications of such state-level policies.

Our analysis showed that simply becoming a rural retirement destination at a particular point in time does not guarantee that a county will be able to attract older persons continuously over a number of decades. In fact, pathways into and out of the rural retirement destination category are complex and vary markedly over time. While 584 nonmetropolitan counties qualified as RRDs at one time or another since 1980, only 111 counties have been in this category continuously since rural retirement destinations were first delineated. Many counties have entered and exited the category during the 25 years that this phenomenon has been studied by social scientists and policy analysts. Interestingly, exit from the category is not necessarily permanent. The ability to attract older migrants at high levels tends to vary over time. For example, of the 154 counties that entered the RRD category between 1990 and 2000, half were RRDs in 1980 but had lost this status during the next decade. Once a rural county begins to attract older persons at a relatively high rate, older in-migration tends to become a self-maintaining process over time. Apparently, relationships between origins and destinations become established and are kept alive by on-going social network connections. Hence, a temporary decline in older in-migration does not mean that a rural county is totally out of the "retirement community business."

Since retirement destinations tend to grow more rapidly than other rural counties, and since more rapidly growing nonmetropolitan counties are more likely to be converted from nonmetropolitan to metropolitan status, we examined the possibility that exit from the RRD category is associated with metropolitan reclassification. Our analysis indicated that metropolitan reclassification accounted for very *little* of the dropout between 1980 and 1990, but almost half of the loss during 1990–2000. Interestingly, we also determined that of the 65 nonmetropolitan retirement destinations that became metropolitan counties during the 1990s, 20 still attracted older in-migrants at a sufficient rate to retain their retirement destination status. In other words, they were reclassified from rural to urban retirement destinations. In contrast, the remaining 45 RRDs that exited the category because they became metropolitan also experienced a decline in older migration below the 15 percent growth threshold required to maintain retirement destination status. These counties would have dropped out of the RRD category even if they had not been reclassified as metropolitan.

Our multivariate analysis of why some nonmetropolitan counties are more likely to be RRDs than others advances understanding of this question beyond previous research based on reports of destination choice by recent in-movers. While we believe that reports by in-migrants are credible and important, they are subject to the usual shortcomings of survey research that is used to explain past decisions and behavior.[15] Moreover, surveys typically ask respondents to report the most important reason they chose a particular location, or they are asked to identify several reasons which are not then weighted in terms of their relative importance. While such analyses are illuminating, they focus somewhat narrowly on one or two factors. In contrast, our analysis simultaneously considered the effects of four domains of variables on the likelihood that a nonmetropolitan county would be a RRD and that counties would enter or exit the category. Some of the factors included in our modeling were suggested by survey responses of older in-migrants, while others were derived from more general research and theory on population aging and regional development.

Our analysis of why some counties are more likely than others to be RRDs in 2000 showed that although RRDs tended to start smaller than non-RRD counties in the 1970s, they have grown and are still growing rapidly. This leads us to wonder whether relatively small size, one of the most frequent reasons given by in-migrants to explain their residential choice, will continue to be associated with RRD status in the future. Perhaps the perceived benefits of small size become unimportant once a county has been attracting older migrants for a considerable period of time. Small size may be important early in the process of retirement community development because it may facilitate in-migrants becoming socially integrated. However, small population size seems to become less important once the process of older in-movement has been going on for some time and previous in-movers can provide entrée for later arrivals to the community's institutions, organizations and volunteer apparatus. Consistent with this observation, our multivariate analysis showed that more highly urbanized counties were less likely than less urbanized counties to lose their RRD status during 1990–2000.

RRD county economies are highly dependent on recreation and tourism employment, but they are not more likely to have a higher score on the natural amenity scale that includes various aspects of climate, topography and water resources. It might be assumed that this is due to inter-correlation between natural amenities and the measure of recreation dependence, but the correlation between these two variables is only .26. We are not sure why amenities do not have a strong positive association with older in-migration. We think it may be because attitudes toward amenities are so subjective that what is attractive to one in-migrant may be counterbalanced by negative attitudes toward this same attribute among other persons. Clearly this issue is not resolved and more research is needed. Finally, even though only 111 counties have been RRDs continuously since 1980, previous status as a RRD is strongly associated with being a RRD in 2000.

Our analysis of factors associated with joining a RRD category between 1990 and 2000, are quite consistent with the above results. Rate of population change, initial levels and change in percent 65 and older and dependence on recreation and tourism are the only significant factors that predict whether a non-RRD county will enter the category, while dependence on farming is the only factor that reduces the likelihood of joining the RRD category. All of the remaining variables were statistically insignificant.[16] Accordingly, while it makes sense that growing and aging counties with a high degree of dependence on recreation and tourism are more likely than other areas to attract older persons and become retirement destinations, this doesn't really shed much light on the dynamics of the process. In addition, the direction of causation between population growth, population aging and becoming a RRD is not clear. One could just as easily argue that being a RRD results in faster growth and more rapid aging. We cannot explain why presence of a college, increased representation of persons with college education, adjacency to a SMSA and other factors that would seem to make an area attractive to older persons fail to predict entry into the RRD category.

Our analysis does a better job of explaining why counties drop out of the RRD status. Growing places, more highly urbanized places, older age structure places, places that are aging relatively rapidly, college towns and places with increasing educational attainment are *less* likely to drop out of the RRD category between 1990 and 2000. In contrast, counties with higher migration rates from out of state and those with relatively rapidly growing minority percentages are more likely to drop from the category. It appears that older in-movers avoid places which are becoming increasingly diverse with respect to race and ethnicity. Finally, given its positive impact on the likelihood of being and becoming a RRD, we were surprised that a relatively low level of dependence on recreation and tourism is not a significant predictor of dropping out of the category.

The multivariate analysis has shed some light on the question of why some nonmetropolitan counties are more likely to attract older persons, but it is simply a first step toward an integrated answer to this important question. Moreover, like most research, it identifies new questions for future investigation. For example, we recommend that future research should attempt to reconcile differences between county-level ecological analyses, such as this one, and survey results where in-migrants report their reasons for destination choice. Why, for example, is not adjacency to a SMSA, or presence of an institution of higher learning associated with becoming or being a RRD? Why are the results for natural amenities so confusing and often so counter-intuitive? Future analyses should re-examine why some of the variables we expected to predict RRD status failed to do so.

NOTES

1. As will be discussed later in the chapter, we deleted 3 RRDs from the category for various statistical reasons, hence using a total of 274 RRD counties in our analyses.
2. These reclassifications used the same rules, but data from the latest decennial census. As will be discussed later in this chapter, we date the 1979, 1989, and 2004 reclassifications as 1980, 1990 and 2000 respectively for ease of presentation.
3. Migration is indicated by changing one's "usual residence" from one county to another. See Voss et al. (2004) for a discussion of the computation of county-specific net migration by age.
4. It should also be noted that in the 2000 period, for the first time, the ERS county typologies were developed for all counties, allowing both metropolitan and nonmetropolitan counties to be retirement migration destinations. This chapter, however, mostly focuses on the nonmetropolitan retirement migration destination counties.
5. We thank John Cromartie, Kenneth Johnson, David McGranahan, and Max Pfeffer for providing the county level data that we used in creating the database for this analysis.
6. It should also be remembered that among non-RRD counties many of the most rapidly growing ones were switched to the metropolitan category leaving a smaller residual behind.
7. As indicated by the percent of population aged 25 and older that has completed at least some college.
8. The natural amenity scale includes measures of winter and summer temperature, winter sun, topographic variation and water area. It ranges from −6.4 signifying unattractive amenities to 11.17 signifying attractive amenities. Approximately half of nonmetropolitan counties have negative scores (McGranahan, 1999)
9. We looked at both 4 year and less than 4 year institutions in both the public and private category, but the RRD vs. non-RRD comparison was similar. Because of this, we combined them into one community attribute, *presence of a college*, in this comparative profile and in the multivariate analysis that follows.
10. Coefficients in Tables 3.5, 3.6 and 3.7 are odds ratios which are exponentiated logistic regression coefficients. The odds ratio is an indicator of the change in odds resulting from a one-unit change in a predictor. Variables with odds ratios greater than 1.0 increase the probability of being a RRD, while those with odds ratios below 1.0 reduce the chances. Odds ratios tell whether particular variables increase or decrease the chances that a county will be in the category of interest, holding other factors constant. In the analysis displayed in Table 3.5, for example, this means being a RRD in 2000.
11. The natural amenities scale is a measure of the physical characteristics of a county. The scale was constructed by combining six measures of climate, typography and

water area that reflect environmental qualities. These measures are warm winter, winter sun, temperate summer, low summer humidity, topographic variation and water area. (ERS 2004)

12. We are aware that controlling for the South makes the model reflective of only non-South counties. However, models were run with and without the binary control for the South to look for significant differences in coefficients and significance levels. Our analysis indicates that both versions of the model resulted in almost identical results. Therefore, not only does being in the South not significantly affect the probability of being a RRD, it also shows that the process of being a RRD is similar in the South and non-South counties.

13. Of the 65 RRD counties that dropped out between 1990 and 2000, 26 are located in four states: California (8), Florida (5), North Carolina (6), and Texas (7).

14. Only new nonmetropolitan retirement destinations are examined in this section, even though some metropolitan counties also attracted 15 percent or higher net migration at ages 65 and older for the first time during the 1990s.

15. Survey accounts of past behavior are affected by rationalization, imperfect memory and regression to the mean (Campbell and Stanley 1963)

16. Except employment rate, but as discussed in the text we believe this to be a statistical result not a substantive one.

REFERENCES

Brown, D.L., and Mauer, W. (2005 September). *Losing and gaining metropolitan status: So what?* Paper presented at the annual meeting of the Multi-State Committee on Rural Population Change, Las Vegas. Available as *CaRDI Report* No. 4. at: www.cardi.cornell.edu.

Brown, D.L., Cromartie, J., and Kulcsar, L. (2004). Micropolitan areas and the measurement of American urbanization. *Population Research and Policy Review, 23*(4), 399–418.

Campbell, D., and Stanley, J. (1963). *Experimental and quasi-experimental designs for research.* Chicago: Rand McNally.

Cook, P., and Mizer, K. (1994). *The revised ERS county typology: An overview* (Rural Development Research Report No. 89). Washington, D.C.: U.S. Department of Agriculture, Economic Research Service.

Economic Research Service, U.S. Department of Agriculture. (2004). *Data sets: Natural amenities scale.* Retrieved March 2006, from http://www.ers.usda.gov/Data/NaturalAmenities/.

Economic Research Service, U.S. Department of Agriculture. (2004). *Rural America at a glance.* Retrieved February 2004, from http://www.ers.usda.gov/publications/rdrr97%2D1/

Economic Research Service, U.S. Department of Agriculture. (2004). *Data Sets: County Typology Codes.* Retrieved March 2007, from http://www.ers.usda.gov/Data/TypologyCodes/

Fuguitt, G., and Beale, C. (1993). The changing concentration of the older nonmetropolitan population 1960–1990. *Journal of Gerontology: Social Sciences, 48*(6), S278–S288.

Johnson, K., and Beale, C. (2002). Non-metro recreation counties: Their identification and rapid growth. *Rural America, 17*(4), 12–19.

Johnson, K., and Cromartie, J. (2006). The rural rebound and its aftermath: Changing demographic dynamics and regional contrasts. In W. Kandel and D. L. Brown (Eds.), *Population change and rural society* (pp. 25–50). Dordrecht: Springer.

Litwak, E., & Longino, C. (1987). Migration patterns among the elderly: A developmental perspective. *Gerontologist, 27*(3), 266–272.

McGranahan, D. (1999). *Natural amenities drive rural population change* (Agriculture Economic Report No. 781). Washington, D.C.: U.S. Department of Agriculture, Economic Research Service.

Nelson, P., Nicholson, J., and Stege, H. (2004). The baby boom and nonmetropolitan population change, 1975–1990. *Growth and Change, 35*(4), 525–544.

Plane, D., and Rogerson, P. (1991). Tracking the baby boom, the baby bust, and the echo generation: How age composition regulates U.S. migration. *The Professional Geographer, 43*(4), 416–430.

Ross, P., and Green, B. (1985). *Procedures for developing a policy-oriented classification of nonmetropolitan counties* (Staff Report AGES850308). Washington, D.C.: U.S. Department of Agriculture, Economic Research Service.

Voss, P., McNiven, S., Hammer, R., Johnson, K., and Fuguitt, G. *County-specific net migration by five-year age groups, Hispanic origin, race and sex 1990–2000* (Working Paper No. 2004–24). Madison, WI: Center for Demography and Ecology, University of Wisconsin-Madison

CHAPTER 4

WHO MOVES TO RURAL RETIREMENT COMMUNITIES, AND WHY DO THEY MOVE THERE?

INTRODUCTION

This chapter examines the characteristics of older persons who move to rural retirement destinations (RRDs). We ask the following questions: (1) What kinds of areas do older in-migrants to rural retirement destinations move from? (2) What is the socio-economic profile of older in-migrants to rural retirement destinations, and how do their socio-economic characteristics compare with longer-term older residents of these same communities? and (3) What reasons do older in-migrants give for deciding to leave their previous residences and choosing their new rural retirement communities? Taken together these analyses provide a comparative profile of older residents of rural retirement communities and insights into the decision making processes that result in their residential relocation to rural areas at age 60 or older.

Migration Selectivity

Migration, including that of older persons into rural retirement destinations, is a selective process. Accordingly, it is important to know who is moving into RRDs and how they compare with longer-term residents of these same locations. Understanding the selectivity of migration provides insights into migration's impacts and, since some types of persons are more likely to move than others, examining selectivity provides information about the determinants of migration streams (Walters, 2000). Research has shown that some persons are more likely to migrate than others. For example, younger persons and persons with higher levels of education, human capital and financial resources are more likely to move than older persons and those with more limited resources (Long, 1987). These selectivities make sense because migration is generally considered to be an investment in human capital. Older people have fewer years of life remaining during which to reap the returns from migrating and thereby to justify its costs and risks. More highly educated and occupationally skilled and experienced persons are better able to take advantage of economic opportunities in new labor markets compared with persons with limited human capital. Additionally, individuals with greater financial resources are better positioned to overcome obstacles between their current residence and their destination, thus reducing the risks associated with residential relocation.

Migration can also be thought of as an event in one's life course. Younger persons are more likely to move than older persons because they are beginning careers, completing school and starting families. These life course transitions can result in a need to move (for example, to obtain more space when children are born or when one's job is relocated), and/or with accumulating or losing resources that facilitate or constrain actualizing residential preferences. Older age itself includes a number of important life course events that can be associated with residential relocation. While older persons are much less likely to move than persons in the working and family formation ages, some do move. For these individuals, their mobility is typically associated with their transition from the workforce to retirement and as they advance in age with declining health, social relationships and financial resources. Moreover, destination choice is not a chance occurrence. Migration, including that of older persons, is motivated and steered by social networks (Massey, 1990). Experiences in particular places earlier in life and/or social ties with longer-term residents increase the chances that older migrants will choose certain retirement destinations over others. In other words, destination choice is partly contingent on pre-existing social ties, not simply on the availability of attractive amenities, supportive services and social infrastructure.

While this book focuses on migration of older persons to rural retirement destinations, age and socio-economic selectivities at these older ages are similar to those that characterize urban-rural population movement in general. As Fuguitt and Heaton (1995) have shown, net migration to rural areas has been age-selective throughout recent history. Recent research by Johnson and Cromartie (2006) showed that rural areas lost more young adults than they gained in each decade since 1950, but they gained more persons aged 50 and older than they lost during the majority of this period (see Figure 2.1). Most notably, their research shows that the rural net in-migration rate at ages 50 and above was greater during the 1990s than in any previous decade, even though it was positive in earlier periods as well. And, as shown by Fuguitt, Beale and Tordella (2002), the rate of rural in-migration at ages 65 and older exceeded that at younger ages during the 1970s and 1980s, and was nearly as high in the 1990s when the number of potential older migrants was diminished because some of them were born during the Great Depression when birth cohorts were abnormally small. These analyses show that migration to rural areas is highly selective of older persons and has been so for decades. Research has also demonstrated that migration to and from rural areas shows the familiar socio-economic selectivities of migration in general. The most thorough and long-term analysis of this issue was conducted by Fulton, Fuguitt and Gibson in 1997. Studying the selectivity of metropolitan-to-nonmetropolitan migration from 1975 through 1993, the authors showed that, regardless of the direction of this migration stream, whites always predominated. Similarly, persons with high school or higher education were most likely to move in or out of nonmetropolitan areas during these years, depending on the overall direction of migration; and poor persons were least likely to move. Finally, the authors examined metropolitan–nonmetropolitan migration by occupational status and found that white collar workers are most likely to move out of rural areas when the overall migration pattern involved net loss, and most likely to move in when the reverse was true. They concluded that a concern about the erosion of nonmetropolitan human resources was justified during the 1980s when nonmetropolitan areas experienced overall net out-migration. Nonmetropolitan areas lost younger persons, more highly educated people and people employed in white collar occupations during this decade. The opposite was true during the 1970s and early 1990s when the direction of net migration favored nonmetropolitan areas.

Migration Decision Making

Migration decisions are made at the household level, not by individuals acting in social isolation. The conventional framework posits a comparative decision-making process in which the positive and negative attributes of one's current residence are weighed against those available in alternative destinations, while also taking into consideration intervening obstacles that constrain movement between one's origin and potential destinations (Lee, 1966). In addition, most researchers realize that the expected benefits resulting from migration occur over a time horizon, and therefore they must be discounted because immediate gains are typically valued more highly than those occurring after the fact (Todaro, 1989). The costs associated with moving include actual expenditures, such as transporting oneself and one's possessions, emotional costs associated with separating from long time relationships, associations and memberships, and the relocation costs that occur after one arrives in a new area. Indirect costs, such as opportunities that will be foregone because one is not present to take advantage of them, are also involved in the migration calculus.

Because migration is based on expectations, it entails substantial risk. Researchers have shown that risk is mitigated to some extent by information about alternative locations. Risk reducing information, however, is not equally available to all potential migrants. It is provided by institutional sources, labor and real estate markets and, perhaps more importantly, social networks that link origin and destination communities (Massey, 1990). Accordingly, in this chapter we analyze the pushes and pulls that older migrants report as contributing to their decisions to move to a RRD. We also examine whether they had experiences in their new communities earlier in their lives and/or pre-existing social connections with longer-term residents.

Since our focus is on migration into rural communities, we need to place older migrants' decision-making behavior in the context of the nation's overall residential preferences. While research on this issue has waned recently, a large body of knowledge on the topic was built up during the 1970s and 80s.[1] Fuguitt and various colleagues conducted a substantial body of research on size of place preferences in the US during this period and found convincing evidence of a deeply held preference for rural and small town life among the American population. This research showed that most persons preferred to live in places of equivalent size to their current residence; but among persons whose current and preferred residence differed, they were more than twice as likely to prefer a smaller than a larger place (Brown, Fuguitt, Heaton and Waseem, 1997). Moreover, research has demonstrated that size of place preferences were stable from the 1970s to the 1990s (Fuguitt and Brown, 1990) and preferences for smaller places persisted even after differential access to larger places was considered (Fuguitt and Zuiches, 1975). In other words, the research is clear that, given the opportunity, many Americans will seriously consider a move from larger to smaller areas. It is unsurprising, therefore, that when persons retire or consider reducing the intensity of their links to the labor force they may consider moving to an area that fulfills life style preferences. Thus, for those who decide to move at this time in their life course, a substantial number are likely to choose smaller and more rural places. Moreover, older migrants have been shown to select their new residences based on earlier experiences as visitors, tourists or prior residents (Cuba, 1989; Glasgow, 1980). They may also choose to join older migrants who moved previously to destination communities where retirement living has become well established.

THE CORNELL RETIREMENT MIGRATION SURVEY

The analysis in this chapter is based on two types of individual level data: telephone surveys and face-to-face interviews with a subset of telephone survey respondents.[2] Survey data are our principal data source. The interviews provide supplementary information that adds depth to the analysis. The telephone survey was expressly designed to provide insight about older in-migrants to rural retirement destinations, examining how they compare with similar aged longer-term residents and their motivations for moving. In this chapter, we focus on migrants' social and economic characteristics and on reported reasons for leaving previous communities and moving to their current county of residence. In the following chapter, we use these data to examine the process of migrant adaptation and social integration, as well as the effects of social relationships and social integration on the health and well-being of older residents living in RRDs.

Survey Methodology

The Cornell Retirement Migration Survey obtained data from recent migrants aged 60–85 who moved to rural retirement destinations, and from a matched sample of longer-term older residents of these same counties. The survey's first wave was conducted in the fall of 2002. Respondents were re-contacted in spring/summer 2005. As indicated in Chapter 2, rural retirement destinations are defined as entire counties that experienced 15 percent or higher increase in population at age 60 and above from in-migration between 1995–2000 (Economic Research Service-USDA, 2004). From the list of rural retirement destination counties, we purposely selected 14 counties for our survey.[3] If we had selected a random sample of counties in the retirement migration category, the vast majority would have been in the South and Southwest. But because rural retirement migration is much more geographically extensive (see Figure 2.2), we wanted our survey to reflect the diversity of socio-economic and geographic contexts where retirement migration is a well established phenomenon.[4] Hence, while our survey is not statistically representative of the older population living in rural retirement destination counties, it does reflect the broad diversity of local conditions in rural areas that attract older migrants at higher than average rates. The geographic location of our survey counties is displayed in Figure 4.1.

The survey was conducted by the Cornell University Survey Research Institute (SRI) using computer-assisted telephone survey techniques.[5] Age-targeted samples for the 14 counties were obtained from a commercial vendor and respondents were screened with respect to residence in a study county,[6] aged 60 to 85 and length of residence in the study county. Migrants were defined as persons who had lived in the respective study counties for five or fewer years. The average length of residence for migrants as of the first survey wave was three years. Longer-term older residents of these same counties had lived there for an average of about 20 years. The two samples contained approximately equal numbers of migrants (368) and longer-term older residents (420). When a contacted household contained more than one person aged 60–85, one older household member was chosen randomly to be the respondent. Each telephone interview lasted approximately 30 minutes.

Figure 4.1. Counties Where Cornell Migration Survey Was Conducted,
2002, 2005

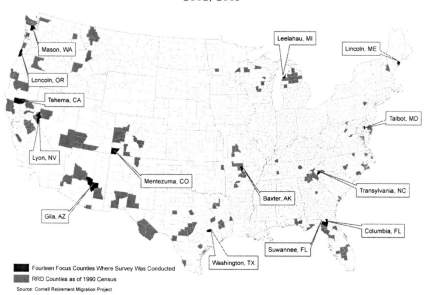

Even in RRDs, in-migration at older ages is a relatively infrequent event, and hence the proportion of migrants in the older population is generally less than the proportion of older persons in the general population. Accordingly, we weighted migrant responses in each of the 14 counties by that county's age-specific migration rates so that in-migrants' records would have appropriate weight when pooled together with data obtained from longer-term older residents. We first produced a weight for each of the 14 counties and then used the population aged 60 and older in each county to combine the individual county weights into an overall weight for the sample. The resulting pooled data represent the population aged 60–85 in the 14 study counties. These pooled data are mostly analyzed in Chapter 5, where we use multivariate analysis to examine factors associated with social participation among older residents of RRDs and the impact of social integration on their health and well-being.

The survey's second panel was conducted three years after the first. Original respondents received yearly post cards bringing them up-to-date about the study and reminding them that we would re-contact them for a second interview in 2005. Additionally, during the first interview we asked respondents where they thought we would be able to contact them in three years hence. Both of these techniques helped us maintain contact with the original respondents. The second wave of the survey was briefer than the first because we already had a full battery of socio-demographic information on the respondents from the first wave. The second wave was focused on examining changes in social relationships, organizational memberships, health and well-being. We were successful in re-interviewing 638 of the original 788 respondents. Attrition between the two waves was due to refusal to be interviewed again (43), inability to re-locate respondent for re-interview (57) and death (50). Thirty-five respondents moved outside of the study county.[7] We compared the socio-economic composition of the wave 1 and wave 2 data and found no statistically significant differences between the two data sets. Accordingly, we were satisfied that attrition between 2002 and 2005 did not introduce bias into the second wave data.

WHO MOVED TO RURAL RETIREMENT DESTINATIONS?

Geographic Origins of In-Migrants

Recent research by Longino and Bradley (2006) found that 74.5 percent of all older migrants between 1995 and 2000 originated in metropolitan areas. They contend that this indicates that metropolitan area retirees are more likely to move than rural older persons, because they are better off at the time they retire and more likely to have been geographically mobile during their working years. This may be true, but since their data only show the prevalence of older migration by place of origin, and not migration rates, it is not possible to conclude that older metropolitan residents are more mobile than older rural persons. That rural older migrants are numerically less numerous is hardly surprising since metropolitan areas account for over three quarters of the nation's total population (Brown, Cromartie and Kulcsar, 2004) and therefore a larger number of metro residents are exposed to the risk of moving. Be that as it may, Longino and Bradley (2006) also show that, regardless of their type of origin, the majority of migrating older persons move to metropolitan areas. Using data from the 2000 US census, they showed that 74 percent of metropolitan-origin migrants moved to other metropolitan areas, as did 62 percent of rural-origin out-migrants. Accordingly, most retirees who move go to other metropolitan areas, albeit often to their peripheries. However, since we are particularly interested in rural retirement destinations, we believe that these data lend support for the continuing strength of the phenomenon. The data show that twenty-seven percent of rural out-migrants moved to other rural areas, as did 18 percent of metropolitan older movers. While most older migrants, including those who moved to rural areas, originate in metropolitan areas, about a quarter of older migrants come from rural areas.

As is shown in Table 4.1, our survey data are consistent with this picture. About three quarters (73 percent) of in-migrants to the 14 rural study

Table 4.1. Metropolitan or Nonmetropolitan Origin of In-Migrants
to Rural Retirement Destinations, 2002

| | Former Residence | | |
	Metropolitan	Nonmetropolitan	Total
Same State	35%	11%	54%
Different State	37%	16%	46%
Total	73%	27%	100%

Source: Cornell Retirement Migration Survey.

counties moved from metropolitan areas, while 27 percent came from other rural areas. Interestingly, rural-origin in-migrants are more likely to have moved across state lines while urban-origin migrants are more evenly split between same state and inter-state migration. Moving across state lines is an indicator of migration distance and these data suggest that a higher proportion of urban-origin older migrants to RRDs move relatively shorter distances.

In 1995, Glasgow conducted a study to examine service utilization among older in-migrants to RRDs in the Middle Atlantic states. She reported that nine out of ten older in-migrants to the rural retirement destinations in her study region came from nearby metropolitan areas (Glasgow, 1995). How can the 90 percent metropolitan area-origins reported in the earlier study be reconciled with the 75 percent urban-origins reported here? Does this difference tell us anything important about regional diversity in the rural retirement migration trend?

The answers to these questions likely lie in the fact that Glasgow's earlier study was conducted in rural retirement destinations located in the Pocono Mountains, Maryland's eastern shore, the Delaware beaches and New York's Catskill mountains. These areas are all located in close proximity to large metropolitan areas, such as New York, Philadelphia, Baltimore, Richmond and Washington, D.C., and are popular weekend and summer locations for urban professionals to own second homes. Thus, it is unsurprising that retirement migration to these destinations was highly dominated by persons from nearby metropolitan areas. Our study, in contrast, was conducted in 14 purposely selected rural retirement destinations scattered throughout the nation. Some of our study communities, like Talbot County, Maryland, drew heavily from nearby metropolitan areas while others, like Baxter County, Arkansas, drew more widely from both urban and rural areas of the Midwest. It is important to recognize diversity in the rural retirement phenomenon and to avoid overall generalizations that cover up important aspects of community differentiation.

As we will discuss in Chapter 6, data obtained in case studies in 4 of our 14 rural retirement survey counties support this view. In Leelanau, Michigan, for example, most of the in-migrants we interviewed came from elsewhere in the Midwest, including many from Detroit, Grand Rapids and other large cities within Michigan itself. In contrast, in Transylvania County, North Carolina, older in-migrants came from throughout the eastern part of the U.S. In fact, many were "half backs," people who had originally moved to Florida from New York, Philadelphia and other large cities in the Northeast but who moved "half way back" to North Carolina in response to Florida's hurricanes and hot, sticky summer weather.

Single or Multiple Residence?

A lot has been written about "snow birds," or older persons who maintain a principal residence in a colder climate where they may have lived during their working life and a second residence in a warmer climate where they spend 3–5 months per year during the late fall and winter (McHugh and Mings, 1991). While the extent of this seasonal migration of older persons is not known, it is thought to be substantial. Krout (1983), for example, estimated that the U.S. snowbird flow exceeds that of older persons making permanent moves to the sunbelt. Seasonal migration to the southern U.S. from Canada is also thought to be significant. Marshall and Longino (1988) estimated that between 225,000 and a half million older Canadians visited Florida each year during the early 1980s. Daciuk and Marshall (1990) reported that Canadian snow birds arrive in Florida in November and return to Canada in April, staying 5–6 months. Moreover, they reported that this was a stable pattern of migration with 75 percent of respondents to their survey reporting that they anticipated returning to Florida on a continuing basis. Some researchers have proposed that temporary migration is a precursor to permanent relocation, but Hogan and Steinnes (1993) cast doubt on this observation by showing that permanent and temporary migration of older persons, while related, are separate phenomena.

If a significant proportion of respondents to our survey are seasonal residents, this would almost certainly depress their inclination to become involved in social relationships and social organizations in the retirement communities. However, 91 percent of our survey respondents reported that the RRD was the "principal residence" where they live for 11–12 months per year. Of those who reported that they spend a lesser amount of time in the community studied, about half reported that they have a second house elsewhere, while slightly fewer reported that they travel and/or visit friends and relatives while away. Like most issues, however, the prevalence of seasonal migration among respondents to our survey varies across the 14 study counties. For example, while almost all of the older in-movers to Lincoln County, Maine, were year round residents, only 80 percent of in-movers to Leelanau County, Michigan, lived there 11–12 months per year.[8] Hence, while seasonal migration among our respondents is not a serious issue in most of the study counties, it could negatively bias our estimate of social integration among older in-migrants in those areas where it is a more common practice.

The overall low prevalence of seasonal residence in our survey compared with previous research may partly result from the fact that the first wave of our survey was conducted in April, after many of the south and south-western snow birds had returned to colder climates. Earlier studies of snowbird migration indicated that many of these persons live in mobile homes and recreational vehicle parks (Martin, Hoppe, Marshall and Daciuk, 1992). However, only 7 percent of respondents to our survey reported living in a mobile home and 86 percent reported that they live in a single family dwelling. Moreover, greater than ninety percent of respondents to our survey own their own homes. Our case study in Gila County, Arizona, suggests that our survey's timing did limit the number of part year residents in the data. We visited Gila County, Arizona, during October, 2006, and a large number of seasonal RV residents had already congregated near Roosevelt Reservoir. In contrast, survey data for this county showed few part-year residents and few residents living in mobile homes. Hence, if we had conducted our survey in this particular county during the winter, we likely would have had more seasonal residents in our sample. This potential bias should be noted when considering the overall generalizability of our findings for RRDs. A study conducted in the winter would have almost certainly included more part-year residents in the warmer states. Be that as it may, most respondents to our survey are permanent residents, which means that the majority of them have access to social opportunities and good reasons to invest in social and community relationships.

Selectivity of Migration to Rural Retirement Destinations

The data in Table 4.2 compare the socio-economic and demographic characteristics of migrants and longer-term older residents of rural retirement destinations as reported in the first wave of the survey.[9] These data confirm the selectivities reported in previous research, which indicated that migrants to rural retirement destinations are positively selected (Glasgow, 1995; Longino, 1990). In-migrants are younger and more likely to be male and married living with their spouse/partner (see Figure 4.2). They are also more highly educated, although more than one third of longer-term older residents also have post-secondary education, which is higher than we expected.

In-migrants are more likely to have been retired at some point than longer-term older residents, but again the difference, about 8 percentage points, is not great. The vast majority of both populations are no longer employed (see

Table 4.2. Comparative Profile of Older In-migrants vs. Longer-term Older Residents in Same Counties, 2002

	Migrants	Non-migrants	Total Population Age 60+[a]
Demographic and socioeconomic characteristics			
Currently married/partnered	74.9	68.4	70.3
Median age (years)	67.0	71.0	70.0
70 or older	36.8	56.6	52.8
Female	48.6	66.2	63.1
Less than high			
school graduate	3.5	8.1	7.4
High school graduate	23.9	27.5	27.5
College or post graduate	43.5	34.7	35.5
Ever retired	91.0	82.7	84.2
Currently working			
for pay	36.9	34.8	34.3
Median years in county	3.0	22.0	15.0

Source: Cornell Retirement Migration Survey.
[a] Weighted.

Figure 4.3). It is interesting, however, that while over 80 percent of respondents report having retired at some point in their life course, more than a third of both migrants and longer-term older residents report that they are currently working for pay. Some of these persons are labor force re-entrants, others have worked continuously without ever retiring and many are working part-time (see Figure 4.4). The importance of current earnings, while not trivial, should not be over-emphasized. Less than 15 percent of respondents reported that earnings from current work are an important part of their annual household income. In contrast, more than half of in-migrants and longer-term older respondents reported that Social Security was a very important source of current income. Additionally, over half of in-migrants and about 40 percent of longer-term residents depend heavily on pensions from former employment and savings and investments (data not shown). In summary, while these data show the expected positive selectivity of in-migrants, differences between older in-migrants and longer-term older residents of RRDs are less than we expected. This higher than expected level of socio-economic status among longer-term residents may reflect the fact that many rural retirement communities have been attracting older migrants for a considerable time, and hence many of the longer-term residents may themselves

Figure 4.2. Retired Doctor Who is Still Active in Relief Work in Turkey

have been in-migrants in earlier years. In fact, only 13 percent (47) of longer-term residents were born in the rural retirement communities, and although they have lived in their current communities for an average of 22 years, many have moved in more recently.

To Move, and If So, Where?

Migration decision making is a multi-step process, and each step in the process is influenced by a different set of factors. Wiseman (1980) modified the conventional push–pull migration framework (Lee, 1966) to explain why older people move. He observed that each step in the process is regulated by the fit between persons and their natural, social and economic environments. As various pushes and pulls act on persons, they either trigger moves or result in efforts to adjust the present location, thereby diminishing environmental stress. Initial steps in the migration process are focused on ascertaining the need and feasibility of moving or staying. Once potential migrants have decided that moving is in their best interest, they select a destination from among alternatives.

Figure 4.3. In-migrants Typically Arrive as Couples

Research on migration decision making typically emphasizes economic factors, but this seems unrealistic in the case of older movers whose "moorings" to the labor market have weakened (Moon, 1995). Accordingly, the role of non-economic determinants of geographic mobility takes center stage in research on migration during later life. This is consistent with a more general contention that non-economic factors should not be subordinated in studies of migration decision making, regardless of age. For example, Halfacre (2004) argues that examining the non-economic worlds of migration reveals migrants' cultural aspirations and permits one to understand geographic mobility as an aspect of personal biography and life course.

While migration decision making was not the major focus of this research, our survey provides insight into the features older in-migrants to RRDs considered problematic in their previous communities, as well as the factors and social relationships that helped to steer them to rural retirement destinations. Because all of our respondents have already moved, we know that the decision to move prevailed over staying put. Knowing the kinds of factors they perceived as problematic in their previous places of residence and those that attracted them to RRDs provides a context within which to examine older migrants' ease or difficulty in adapting to their new residences.

Figure 4.4. In-migrant Who Owns a Home Repair Business

        Migration motives can be analytically separated into a number of
distinct values. De Jong and his colleagues (1995) identified six motives (or
values) associated with migration among persons aged 70 and older: health,
affiliation, economic security, comfort, functional independence and getting on
with life after a crisis. However, since our sample's migrants were aged 60–85
at the time of the first wave, with an average age of 68.1 years, functional
independence and getting on with life after a crisis are probably less salient
than was true in De Jong's research where migrants were 70 or more years
of age. Rather, we find the categorization of migration motives proposed by
Glasgow (1980) in her analysis of retirement migration to the Midwest to
be more useful for organizing our analysis of reasons for leaving the prior
residence and choosing a destination. Glasgow organized these reasons into
five categories including: employment, environment, ties, retirement and other.[10]
These reasons include community attributes that respondents cite as attractive
in their new locations and, also, whether they had pre-existing social and/or
economic relationships in their new communities. We have retained Glasgow's
logic, but elaborated her framework into the following seven categories: life
course changes (including retirement), community attributes (including urban
externalities and rural amenities), economic security, environment (including

climate and weather), family, prior social ties and other. Respondents were asked open-ended questions which were then coded into the above categories. Their responses are displayed in Table 4.3. The questions asked in our survey are listed below. In the analysis that follows, actual quotations from these open-ended questions are inserted into the text to illustrate and ground major points.

- What was it about your previous place of residence that influenced your decision to move here?[11]
- Why did you choose to move to your current county?

Reasons for Leaving

Community amenities were mentioned by almost one quarter of migrants as the main reason for leaving their prior residence. These amenities also include perceived negative aspects of previous places of residence. Respondents identified traffic, crowding, congestion, crime, taxes and air pollution as reasons for moving. Population size was also important. One respondent indicated that "over-population has gotten out of hand." Another observed that "Phoenix has grown too big," while several persons observed that "we want to live in a smaller town." Several respondents mentioned changes in ethnicity as being problematic. One respondent observed that the "influence of Hispanics" had become too great. Another person said that there were "too many people and they were the wrong type of people" in the previous residence. While not frequent, several of the

Table 4.3. *Main Reasons for Leaving Previous Residence and Choosing Current Residence, 2002*

	Main Reason for Leaving Previous Residence (Percent)	Main Reason for Choosing Destination (Percent)
Life course transition	18.6	3.6
Community attributes [a]	24.2	25.5
Natural environment	18.6	20.3
Family relationships	16.8	23.2
Economic security	14.3	12.7
Prior experience in area [b]	5.0	8.2
Miscellaneous	2.5	6.5
Number of responses	161	306

Source: Cornell Retirement Migration Survey, Wave 1.
[a] Includes both big city externalities and positive and negative aspects of rurality.
[b] Not including family ties.

persons we interviewed during the case studies also mentioned racial and ethnic issues as being among the factors that persuaded them to leave their previous communities.

It is worth remembering that more than one quarter of in-movers to rural retirement destinations came from other rural areas. For these persons, the new community typically was larger, with a more highly urbanized environment. These respondents criticized the smallness and rurality of their previous communities. One respondent remarked that "it was too remote, too far from shopping," while another commented that they were "too far out in the country before."

Almost one in five respondents indicated that they left their previous residence because of *life course transitions*. Of those who identified life course changes as their main reason for leaving, about one third indicated that retirement was the precipitating event. The other two-thirds were a mixed bag including divorce, death of a spouse, simply wanting a change and various health-related reasons, including several associated with their declining ability to drive an automobile. Ironically, research on personal transportation among older populations, including the four case studies we conducted for this book, indicates that older persons in rural areas are much more dependent on the private automobile than is true in urban locales (Glasgow, 2000).

Negative *environmental characteristics* were equally as important as life course changes in motivating respondents to leave their prior residences. Most environmentally-oriented respondents identified weather-related issues as being problematic in their previous residences. Snow and heat were the most typical weather-related reasons. Other environmentally-related reasons for moving included not being near the coast and/or the mountains, and to gain distance from earthquakes and pollution.

Family is the next most important factor motivating respondents to leave their previous residences. Respondents typically stated that their children, grandchildren or other relatives did not live near their previous residences. Several indicated that their parents or in-laws lived elsewhere and needed their help. Because many respondents indicated that children or other relatives lived in the rural retirement destination prior to their moving in, it is difficult to separate family as a push factor ("We had no family there") from a factor that pulls migrants to their new community ("We're getting older, and we need to be closer to our children"). This motivation to move closer to one's children is consistent with previous research (Silverstein and Angelelli, 1998).

Economic security is the final reason given by a substantial number of respondents as motivating them to leave their previous residence. Taxes and high costs of living were frequently mentioned; but for those identifying economics as their main reason, half indicated that employment-related reasons were paramount. A number of persons were transferred to the rural retirement community and several had pre-existing businesses there. These responses are consistent with what we observed during our case study interviews. We were surprised by the number of older persons who were employed, some part-time and some full-time, in the four retirement destinations we visited. In other words, moving to a rural retirement destination at age 60 or older does not necessarily indicate that the in-migrant is retired. In fact, our survey showed that 35.8 percent of respondents worked for pay at the time of the initial survey. Moreover, being retired when one moves is no guarantee that one will not re-enter the labor market. During the case studies we interviewed retirees who had opened businesses after moving and some persons who had transitioned from volunteer work to part-time or full-time paid jobs.

About 5 percent of respondents mentioned *prior ties* as reasons for moving from their previous communities, but these reasons seem more like pulls than pushes. Most persons who fall into this category indicated that they had left the previous community because they owned land or a house in the destination prior to moving there.

Reasons for Choosing Current Community

Community attributes were mentioned most frequently as the main reason for choosing the particular rural retirement destination. While this category is somewhat ambiguous, respondents often equated it with notions about "rurality." Some in-migrants observed that they were attracted by the place's "slower pace of life," "small town atmosphere," "small town community feel," "laid back feel," "quiet feel" and "convenience away from the big city." One respondent commented that he moved in order to "get into the country," while another commented that he wanted to raise cows. A number of respondents commented that they chose the place because of its "quality of life." Some respondents were attracted by the area's housing stock, while others responded positively to outdoor recreational opportunities, golfing in particular. Our analysis shows that preferences for rural living are shaped by a complex mix of pro-rural and anti-urban attitudes. Older in-movers to rural retirement destinations were often "pushed" by urban externalities and "pulled" towards rural amenities when choosing where to go.

Environment is both a push and a pull for older in-movers to rural retirement destinations. While weather was a major aspect of environment associated with leaving one's previous community, it is a much less important factor in choosing a destination. In choosing a new community, respondents were likely to mention scenery and the natural landscape. One respondent stated that he "wanted to live in a rural area near the coast and mountains." Another mentioned the destination's "seaside charm." Many of the environmentally-oriented in-migrants indicated that they were attracted by the area's outdoor recreational opportunities. Quite a few respondents commented that they wanted to live where they could hunt and fish. Others commented that they owned horses and wanted to live where boarding horses was affordable and convenient.

Family was the "other" main factor drawing older in-migrants to rural retirement destinations. Almost half (53/119) of respondents who identified family as a pull factor indicated that one or more of their children lived in the rural retirement destination prior to their moving there themselves. Most respondents made simple declaratory statements such as "to be near our daughter" which do not reveal the nature of their attraction. Others revealed that being closer to children was important for reasons in addition to simply being able to visit more frequently. Several respondents indicated that they moved closer to their children in order to visit grandchildren more often. Others indicated that moving closer to their children was at least partly motivated by the expectation that they might need care as they aged. For example, one person stated that "it is important to be closer to my daughter because I'm older now" which revealed a realization that assistance might be needed in the near future. Another respondent indicated that moving was "my daughter's decision," which questions the respondent's degree of autonomy and personal agency. After children and grandchildren, siblings were the most frequently mentioned relatives drawing older persons to rural retirement destinations. In addition, a couple of respondents indicated that they had moved to a particular place in order to care for aging parents or in-laws. One man indicated that he and his wife had moved to the community so that his wife and her brother could care for their mother. Another observed that she and her husband would now be able "to care for her aging in-laws." While family is a stronger factor in destination choice than in deciding to leave one's prior residence, the dynamics are similar. Older persons leave areas where they have few or no family members and choose new communities where children and/or siblings already reside. As will be shown in the next chapter, over one third of all in-movers had at least one child living within one a half hour drive of their new community and only slightly fewer had other relatives living within one hour. In other words, while retirement migration is often motivated by environmental and community amenities, it also contributes to family reunification.

About one in ten older in-migrants indicated that they moved to their new community for *economic reasons*. In general, these persons indicated that they chose their new community because it was affordable and/or had a relatively low cost of living. Housing affordability was a key factor for about a quarter of these persons. One respondent observed that he and his wife needed a "lower mortgage" now that they were on a fixed income. Another man stated that he "got a good deal on this place." Taxes were also mentioned as a reason for destination choice. Nevertheless, affordability was not the only economic reason drawing older persons to rural retirement destinations.

Over one-third of respondents reported that they worked for pay at the time of the first survey, and this appears to be one of the reasons why they chose a particular community. One respondent indicated that he got a good job opportunity that was "near to the city." Several respondents said that they were moving to work in pre-existing family businesses. It is interesting, however, that a much higher percentage of respondents reported working for pay than indicated that job related reasons drew them to the community. This is true whether we look at their main reason for leaving or any of their top three reasons. One respondent's statement is particularly revealing. He indicated that "I took a job, but I only wanted to live on the coast." Moreover, our case studies in four of the fourteen retirement destinations indicated that pathways to and from paid work are complex for in-movers. Some move as a form of pre-retirement and plan to switch from full- to part-time work after they arrive. Some previously retired in-migrants re-enter the paid work force after moving. Sometimes this is due to economic necessity, while in other cases it serves as a way to fill their time and make social connections. It can also occur when they begin to get paid for work they had previously done on a volunteer basis. Some in-migrants are fully employed when they move and retain their full-time jobs, often telecommuting or commuting long distances by car or plane. In addition, we also met a number of in-movers who had established new businesses after moving to a rural retirement community. So, while we characterize this phenomenon as "retirement migration," not all older in-movers are retired.

Prior experience in the rural retirement destination was the main reason 8.2 percent of in-migrants chose to move to their particular community.[12] Prior ties are much more important as a pull factor to their new locale than they are as a push factor from their previous community. Over one third of persons who indicated prior social ties were the main factors that steered them to a particular community owned land and/or a house in the community before moving there. Some had inherited the property, while others had purchased it earlier in their lives. In addition, one out of five persons in this category had lived in the area previously and several were born there (see Figure 4.3).

Many persons who indicated that they were drawn to the community by prior ties became familiar with the area through vacationing. One respondent observed that he had "visited here for 20 years." Another person commented that "I love Maine and I missed it." A married couple we interviewed in one of the case study counties told us that they had rented a house there every summer for 15 years and then built their own place right down the road. Many of the case study respondents we interviewed told us that vacationing in the particular area was a family tradition. They had visited there as children and brought their children to summer there as well. Many build large houses so that they can have family reunions in the summer with 3 and 4 generations participating. Some of the persons we interviewed during the case studies told us they subscribed to the local newspaper so they could keep up with events there during the off season. This kind of continuous connection between vacationers and vacation communities built strong links which contributed to permanent migration in later life. One respondent commented, "I knew a bunch of people in the area."

These findings are consistent with previous research. Glasgow (1980) showed that older migrants to the rural Midwest during the 1970s had extensive prior ties with the destination areas. Many had kin and/or pre-existing friendships in the area. Others owned property there before moving and/or had either vacationed or visited friends or relatives in the area before moving there themselves. Several authors have identified vacationing and/or visiting an area prior to moving there in later life as an important aspect of the retirement process. Marcouiller, Kedzoir and Clendenning (1996) showed that older second home owners in northern Wisconsin learned about the area by vacationing there and/or by having friends and relatives in the area. Cuba (1989) reported that over 90 percent of retirement-age migrants to Cape Cod had previous experience there as vacationers. In fact, many had vacationed on Cape Cod since childhood. Miller and Elliot (2007) conducted surveys and focus groups among retired in-migrants to nine communities in Arkansas, Oklahoma and east Texas; the research showed that scenic beauty, recreational opportunities, climate and low costs were the most important reasons given for choosing to move to these areas. In addition, closeness of family was highly ranked in a number of the areas, which the researchers interpreted as an indicator that persons were returning to areas of previous residence during retirement. Our study, while substantiating the importance of prior ties in the older migrant attraction process, does not show it to be as powerful a motivation as was the case in previous studies. Instead, community, family, environment and economics were mentioned more frequently as the main reason for moving to rural retirement destinations. It should also be noted that our data show that the importance of prior ties as a main reason for destination choice varied dramatically among in-migrants to the 14 RRDs—from 18 percent among persons who moved to Maine, to less than 5 percent among persons who moved to Michigan, Arkansas, Colorado or California. This underscores the need for caution when searching for covering explanations for complex phenomenon such as migration decision making.

CONCLUSIONS

This chapter described older in-migrants to rural retirement destinations and compared these persons to similar-aged longer-term residents. We also examined reasons in-migrants give for leaving their previous locations and reasons that attracted them to their current rural residence. Consistent with previous research, older persons who moved to rural retirement destinations are positively selected. They are younger, more likely to be male, more highly educated and more likely to be living with a spouse/partner. While these differences are important, the data presented in the chapter showed that migrant and non-migrant differences in socio-demographic status are not large. We speculate that many of the longer-term older residents were migrants themselves earlier in their lives. Previous research had contended that older in-movers to rural retirement destinations are mostly from urban areas. Our data also show that three-quarters of in-movers came from metropolitan areas, but a substantial proportion, especially persons migrating across state lines, came from other rural areas. We also found substantial differences among our study's 14 retirement destinations in the percentage of in-movers from metropolitan areas. Accordingly, framing rural retirement migration as an overwhelmingly urban-dominated phenomenon masks important differences among migration streams.

This chapter's most important contribution is the insight it provides into the migration decision-making process. If we can draw any conclusions from the open-ended survey responses, it is that migration decision making involves multiple reasons. While migration during working life is typically motivated by employment-related factors, these were shown to be relatively unimportant to the older in-movers in our study. This is because of their older ages and their relatively high socio-economic status and levels of economic security. The rural retirement destinations we studied in this research are substantially better off financially than other nonmetropolitan counties (see Chapter 2). If we had chosen one or more retirement destinations that typically attract older working class in-movers, cost of living, affordability and availability of jobs might have been more important pull factors. In fact, if we had conducted the interview during the winter when most snowbirds are in residence in our southern and southwestern retirement destinations, it is possible that economic reasons might have been more important.[13]

Respondents to our survey indicated that quality of life factors motivated their decision to leave their previous residences and attracted them to particular destinations. While "quality of life" is an ambiguous term, our respondents often mentioned landscape, community atmosphere, weather and outdoor recreation opportunities. Many told us that they were seeking a small town atmosphere. Even some in-migrants who identified economic factors as their main reason for moving to a particular place indicated that they wanted a job near the coast or in the mountains, away from the smog or snow. In other words, not only do different persons have different reasons for moving, but many persons have multiple reasons and some of these are contingent on each other.

These findings are consistent with the predominant theory of older migration, which focuses on the role of amenities in steering retirees to rural destinations during the initial move after retirement (Litwak and Longino, 1987). However, over a quarter of respondents to our survey reported that family was the main reason for their move. In other words, while a substantial proportion of older persons who move to rural retirement destinations do so to satisfy quality of life and amenity preferences, many also seem to move as a way of re-unifying their families. To some extent the motive behind this preference is to be able to visit relatives more frequently and to have more contact with grandchildren (among those with grandchildren). In addition, it seems that older persons are realistic in anticipating their needs for assistance as they age. Being close to children and other relatives is one way to insure greater access to helper networks that contribute to maintaining independent living for the longest possible time.

Prior social ties in the destination were not a major reason drawing older in-migrants to most of the 14 rural retirement destinations in our study. However, in four of the destinations, prior ties exceeded 15 percent and were one of the top three reasons for choosing to live in the particular community. In these cases, respondents described extremely strong multi-generational ties with the retirement community, sometimes featuring long time land ownership and previous residence in the area.

The next chapter continues at the individual level and examines the degree and nature of social participation among in-migrants to rural retirement destinations. We attempt to explain why some in-migrants to rural retirement destinations are more likely to become socially integrated than others, and we explore the outcomes of social integration with respect to migrants' health and functional abilities.

Retirement migration is not a behavior that participants take lightly. Migrants realize that pulling up stakes after living in a community for many years poses the challenge of becoming socially involved in their new location. Accordingly, once the decision to move has been reached, they carefully consider alternative destinations evaluating environmental, community, familial and economic aspects. Because older in-migrants are positively selected compared with the longer-term population, they are a potential resource for the destination community. We examine their community level impact in Chapter 6.

NOTES

1. One notable exception is Von Reichert's (2006) study of community evaluation and migration intentions in the Northern Great Plains.
2. Since the telephone survey provides the main source of data for this chapter, we detail its methodology here. Case studies provide supplementary information which adds depth to the survey analysis. Accordingly, the case study methodology is detailed in Chapter 6 where it is the primary source of information used in the analysis.
3. The number of survey counties was dictated by our budget because we wanted to interview approximately 60 respondents per county.
4. It should be noted that based on the results of the 2000 US Census, there are currently 277 rural retirement destination counties. These counties were announced in 2004 by USDA. We designed our survey in 2002 prior to the re-classification, however, at which time there were only 189 rural retirement destination counties (based on the results of the 1990 census). All but one of the 14 survey counties we selected from the list of rural retirement destinations as of 1990 are still identified as retirement destinations in 2004. Tehama County, California, however, has been deleted from the category because it no longer has the requisite 15 percent or higher in-migration at age 60+. However, we have retained the Tehama County respondents in our study because they moved to a legitimate rural retirement destination county at the time our survey was conducted.
5. The survey was managed by Yasamin Miller, SRI's director.
6. Since telephone area codes often span county lines, some persons on the list could live outside of a study county.
7. Persons who moved outside of study counties were re-interviewed, but we do not include this information in this chapter.
8. This issue is important to local planners in Leelanau as they attempt to manage growth in the county. Their 2000 publication, "Seasonal Population: The Leelanau General Plan" showed that 36 percent of housing units in the county were seasonal in 1999. (Leelanau Technical Assistance Plan, 2000).
9. The data in columns 1 and 2 of Table 4.2 are un-weighted. The total column which pools data for in-migrants and non-migrants is weighted.
10. These reasons are consistent with several other studies of retirement migration including Rex (2002) and Sofranko, Fliegel and Glasgow (1982).
11. Respondents were asked for both their main reason and top three reasons for leaving their previous residence and choosing a new community. The distribution of all reasons across the seven categories is very similar to the distribution of main reasons. Accordingly, we only examine main reasons in this chapter.
12. Having prior ties was ranked higher than economics when we examined all three reasons for destination selection, but it was still far behind community, family, and environment. Thirteen percent of respondents identified prior ties among their top three reasons, but only 8 percent identified prior ties as their main reason.
13. Calvin Beale pointed this out to us. He recommended that we visit Sierra County New Mexico which attracts a high number of modest income snowbirds and has numerous RV parks.

REFERENCES

Brown, D., Cromartie, J., and Kulcsar, L. (2004). Micropolitan areas and the measurement of American urbanization. *Population Research and Policy Review, 23*(4), 399–418.

Brown, D., Fuguitt, G., Heaton, T., and Waseem, S. (1997). Continuities in size of place preferences in the United States, 1972–1992. *Rural Sociology, 62*(4), 408–428.

Cuba, L. (1989). Retiring to vacationland: From visitor to resident. *Generations, 13*(2), 63–67.

Daciuk, J., and Marshall, V. (1990). Health concerns as a deterrent to seasonal migration of elderly Canadians. *Social Indicators Research, 22*(2), 181–197.

De Jong, G., Wilmont, J., Angel, J., and Cornwell, G. (1995). Motive and the geographic mobility of very old Americans. *National Journal of Sociology, 9*(1), 31–57.

Economic Research Service, U.S. Department of Agriculture. (2004). *Data Sets: County Typology Codes.* Retrieved March 2007, from http://www.ers.usda.gov/Data/TypologyCodes/

Fuguitt, G., Beale, C., and Tordella, S. (2002). Recent trends in older population change and migration for nonmetro areas, 1970–2000. *Rural America, 17*(3), 11–19.

Fuguitt, G., and Heaton, T. (1995). The impact of migration on the nonmetropolitan population age structure. *Population Research and Policy Review, 14*(2), 215–232.

Fuguitt, G., and Brown, D. (1990). Residential preferences and population redistribution, 1972–88. *Demography, 27*(4), 589–600.

Fuguitt, G., and Zuiches, J. (1975). Residential preferences and population distribution. *Demography, 12*(4), 491–504.

Fulton, J., Fuguitt, G., and Gibson, R. (1997). Recent changes in metropolitan nonmetropolitan migration streams. *Rural Sociology, 62*(3), 363–384.

Glasgow, N. (2000). Transportation transitions and social integration of nonmetropolitan older persons. In K. Pillemer, P. Moen, E. Wethington and N. Glasgow (Eds.), *Social integration in the second half of life* (pp. 108–131). Baltimore: Johns Hopkins Press.

Glasgow, N. (1995). Retirement migration and the use of services in nonmetropolitan counties. *Rural Sociology, 60*(2), 224–243.

Glasgow, N. (1980). The older metropolitan migrant as a factor in rural population growth. In A. Sofranko and J. Williams (Eds.), *Rebirth of rural America: Rural migration in the Midwest* (pp. 153–169). Ames, Iowa: North Central Regional Center for Rural Development.

Halfacre, K. (2004). A utopian imagination in migration's *terra incognita?* Acknowledging the non-economic worlds of migration decision making. *Population, Space and Place, 10*(3), 239–253.

Hogan, T., and Steinnes, D. (1993). Elderly migration to the sunbelt: Seasonal versus permanent. *The Journal of Applied Gerontology, 12*(2), 246–260.

Johnson, K., and Cromartie, J. (2006). The rural rebound and its aftermath: Changing demographic dynamics and regional contrasts. In W. Kandel and D. L. Brown (Eds.), *Population change and rural society* (pp. 25–50). Dordrecht: Springer.

Krout, J. (1983). Seasonal migration of the elderly. *The Gerontologist, 23*(3), 295–299.

Lee, E. (1966). A theory of migration. *Demography, 3*(1), 47–57.

Leelanau Technical Assistance Plan. (2000). *Seasonal population: The Leelanau general plan.* Leelanau, Michigan: Leelanau County.

Litwak, E., and Longino, C. (1987). Migration patterns among the elderly: A developmental perspective. *The Gerontologist, 25*(3), 266–272.

Long, L. (1987). *Migration and residential mobility in the United States.* New York: Russell Sage Foundation.

Longino, C. (1990). Retirement migration streams: Trends and implications for North Carolina communities. *Journal of Applied Gerontology, 9*(4), 393–404.

Longino, C., and Bradley, D. (2006). A first look at retirement trends in 2000. *The Gerontologist, 43*(6), 904–907.

Marshall, V., and Longino, C. (1988). Older Canadians in Florida: The social networks of international seasonal migrants. *The Journal of Applied Gerontology, 9*(4), 420–432.

Martin, H., Hoppe, S., Marshall, V., and Daciuk, J. (1992). Sociodemographic and health characteristics of Anglophone Canadian and U.S. seasonal snowbirds. *Journal of Aging and Health, 4*(4), 500–513.

Marcouiller, D., Kedzoir, R., and Clendenning, J. (2002). Natural amenity-led development and rural planning. *Journal of Planning Literature, 16*(4), 515–542.

Massey, D. (1990). Social structure, household strategies and the cumulative causation of migration. *Population Index, 56*(1), 3–26.

McHugh, K., and Mings, R. (1991). On the road again: Seasonal migration to a sunbelt metropolis. *Urban Geography, 12*(1), 1–18.

Miller, W., and Elliott, K. (2007). *Arkansas' retirement-age migration: A statewide overview* (Community & Economic Development Report FSCDM2). Little Rock: University of Arkansas Cooperative Extension Service. Retrieved May 2007, from www.uaex.edu

Moon, B. (1995). Paradigms in migration research: Exploring 'moorings' as a schema. *Progress in Human Geography, 19*(4), 504–524.

Silverstein, M., and Angelelli, J. (1998). Older parents' expectations of moving closer to their children. *Journal of Gerontology: Social Sciences, 53*(3), S153–S163.

Sofranko, A., Fliegel, F., and Glasgow, N. (1982). Older urban migrants in rural settings: Problems and prospects. *International Journal of Aging and Human Development, 16*(4), 297–309.

Todaro, M. (1989). Migration and development. In M. Todaro (Ed.), *Economic development in the third world* (pp. 274–287). New York: Longman.

Rex, T. (2002). *Retirement migration in Arizona.* Tucson: Arizona Department of Commerce.

Von Reichert, C. (2006). Community evaluation and migration intensions: The role of attraction and aversion to place on the northern great plains. In W. Kandel and D.L. Brown (Eds.), *Population change and rural society* (pp. 333–356). Dordrecht: Springer.

Walters, W. (2000). Types and patterns of later-life migration. *Geografiska Annaler, 82*(3), 129–147.

Wiseman, R. (1980). Why older people move: Theoretical issues. *Research on Aging, 2*(2), 141–154.

CHAPTER 5

SOCIAL INTEGRATION AND HEALTH OF OLDER IN-MIGRANTS TO RURAL RETIREMENT DESTINATIONS

NINA GLASGOW AND MARIE-JOY ARGUILLAS

INTRODUCTION

This chapter focuses on the social embeddedness of older persons who move to rural retirement destinations (RRDs). It examines the social integration and health status of older persons living in RRDs, and compares the levels of social integration and health among older in-migrants versus non-migrants living in the same RRDs. It examines the process through which older in-migrants become socially integrated in their new communities and uses longitudinal data from a two-wave panel survey to investigate the relationship between social integration and changes in the health status of older persons living in RRDs.

Whether older in-migrants adjust socially with relative ease or with great difficulty after moving to rural retirement destinations is an issue that has received very little attention in the research literature. The social integration and adjustment of older in-migrants in destination communities, however, are important topics of study because previous research demonstrates that well-being among older persons is at least partly contingent on their social relationships (Pillemer, Moen, Wethington and Glasgow, 2000). The very act of moving disrupts older in-migrants' day-to-day social relationships in the origin community, and, even if they had friends, family and/or acquaintances already living in the destination community, newcomers are unlikely to be deeply embedded when they arrive. In other words, long-distance movers, including older in-migrants to RRDs, are potentially vulnerable to becoming socially isolated subsequent to their migration. This chapter examines the extent to which this might be the case and, if so, what its implications are for older migrants' health status and functional ability.

Social integration generally refers to having a strong interpersonal network and also being an active participant in formal organizations and activities in one's community. A large body of literature shows that a high degree of social integration is beneficial to people's physical and mental health as well as their longevity (for example Berkman, 1983; Moen, Dempster-McClain and Williams, 1989, 1992). The conceptual foundation for much of the research on social integration and health is Durkheimian theory, which viewed involvement in social roles and relationships as an indicator of social cohesion with the larger society (Durkheim, 1897/1951). The degree of older in-migrants' social integration has implications for their well-being and, ultimately, whether they are retained in rural retirement destinations. If in-migrants do not adjust well to their destination communities, they may be more likely to move again.

Definition of Social Integration

Researchers have defined and operationalized "social integration" in a variety of ways, and hence it is a concept without an entirely clear definition. One definition of social integration is that it is the degree to which people are embedded in *meaningful* social roles and relationships, the most proximate of which are family, friendships and affiliations with community organizations. The pool from which individuals develop meaningful social roles and relationships is "the entire set of an individual's connections to others in his or her environment" (Pillemer, Moen, Wethington and Glasgow, 2000, p. 8).

Formal social participation refers to organizational memberships and/or community volunteering, while informal social participation refers to individuals' participation in family and friendship networks. Most past studies on the relationship between social integration and health have focused only on people's informal social relationships, or have conflated formal and informal social relationships. However, Young and Glasgow (1998) argued that it is important to distinguish between formal and informal social participation. The distinction is important because formal organizations act as bridging ties to a variety of social venues where resources and information related to health can be obtained, and they are educational environments that enhance an individual's problem-solving ability (Glasgow and Brown, 2006; Young and Glasgow, 1998). Informal social ties, on the other hand, are said to buffer stress caused by social and economic insecurity and to provide emotional support. Informal social relation-ships, however, are often loaded with ambivalent feelings and obligations that have high "opportunity costs" for the individual, which in actuality may induce stress. Participation in formal organizations is almost always voluntary, whereas people are often obligated to maintain their more intimate informal ties. In this study, therefore, we competitively test whether informal or formal social participation is more strongly related to health outcomes. We hypothesize that formal social participation will have a greater positive effect on health than will informal social participation.

Background and Significance

Why should social scientists care about the social integration, health and quality of life of older in-migrants to RRDs? First, migration-related changes in social integration at older ages can affect the process of successful aging. It can transfer persons from communities which provide easy access to meaningful social roles to areas where social participation is difficult to arrange and maintain. Or the opposite can be true. Since successful aging is also productive aging, enhancing social integration among in-migrants is likely to benefit them as well as their new communities. At the community level, it is through changes in age composition that demand shifts associated with population growth or decline are most clearly articulated (Brown and Glasgow, 1991). The age composition of a community imposes requirements and limitations on its institutions and shapes its patterns of consumption, life style and social behavior. This is true in both rural and urban areas, but adaptation to changes in population composition can be especially difficult in rural areas where small size, low-density settlement and/or geographic isolation are limiting forces. Knowing who is moving to an area, their productive activities, their participation in private and public spheres, their needs for services such as health care and recreation, and how long they will remain and be socially involved are important for *judging the positive and negative impacts of retirement in-migration for the destination communities themselves.*

This chapter aims to provide an explanation of how older migrants to RRDs establish social connections in their new locales, enhance understanding of the association between social relationships and individual-level well-being, and illuminate dynamics of migration in the later stages of the life course. Our examination of factors associated with older in-migrants' experiences in retirement destinations can contribute to more informed policy discussions of the challenges and opportunities associated with rural retirement migration (see Chapter 7). After a brief discussion of data and methods, we provide a descriptive and comparative analysis of health status and social participation among older in-migrants and non-migrants. We then present findings from a multivariate analysis of the effects of variability in social integration and a number of control characteristics on changes in older rural residents' health between 2002 and 2005.

Data and Methods

The sampling frame and survey methodology of the Cornell Retirement Migration Survey were explained in detail in Chapter 4, but here we briefly recapitulate the survey methodology. The analysis reported in this chapter primarily uses data from a two-wave panel survey of persons who were aged 60 to 85 in 2002 when first interviewed, and who, to the extent possible, were interviewed again in 2005. Some attrition occurred between the two waves of the survey, thus reducing the total number of respondents from 788 in 2002 to 638 in 2005 (specific characteristics of the 150 wave 1 respondents lost to follow up at time 2 are detailed in Chapter 4). In the descriptive analysis, we use the total sample of 788 respondents as of 2002. In 2005, the number of cases is smaller because of sample attrition. In the multivariate analysis the number of cases is determined by being in the survey in both 2002 and 2005. An important concern in conducting longitudinal research is whether or not samples become more selective over time, but an analysis of the characteristics of the 2002 survey respondents and the 150 respondents who could not be re-interviewed in 2005 showed no statistically significant differences. In addition, 35 wave 1 respondents out-migrated from the RRD county in which they lived in 2002 (19 were classified as in-migrants and 16 as non-migrants in 2002). This further reduced the 2005 total to 603 (282 migrants and 321 non-migrants) as the base for our analysis of 2005 data.[1]

Data obtained from a sample of recent in-migrants to RRDs are compared to a matched sample of longer-term older residents living in the same counties. Older in-migrants were defined as individuals who had lived in a retirement destination community for five or fewer years, and non-migrants (longer-term residents) were defined as those who had lived in a rural retirement county for over five years. In 2002 a set of screening questions was used to establish county of residence, age (between 60 and 85) and length of residence, in order to determine survey eligibility and assign potential respondents to the in-migrant or the non-migrant sample. Using data from the 1990 census, rural retirement destination counties were defined as those having 15 percent or higher increase in population aged 60 and over due to in-movement between 1985–1990.

For illustrative purposes in this chapter, we supplement the survey data with findings from face-to-face interviews that were conducted with a small subset of survey respondents living in four of our original 14 survey study sites. This case study information grounds the survey analysis, and provides a deeper understanding of some of the relationships revealed by the survey. Face-to-face interviews were conducted in Gila County, Arizona, Lincoln County,

Maine, Leelanau County, Michigan and Transylvania County, North Carolina (see Chapter 6 for a detailed discussion of our case study methodology.)

Why Are Some Older Persons More Likely to Participate in Organizations?

In an earlier study we used data from the 2002 wave of the Cornell Retirement Migration Survey to examine factors associated with the likelihood that older residents of RRDs will participate in formal organizations (Glasgow and Brown, 2006). This analysis focused on three types of factors shown in previous research to be associated with participation: (a) socio-demographic status, (b) health and activity limitation, and (c) involvement in close primary social relationships. Prior research has shown that participation is higher among young-old, healthier persons, those with few activity limitations, and persons with more education (Young and Glasgow, 1998). Conversely, participation has been shown to be lower among persons with dense networks of primary group ties and among persons who work outside of the home (Moen, Fields, Quick and Hoffsmeister, 2000).

The analysis found that education was the only factor to have a strong positive effect on the level of formal social participation among both migrants and non-migrants. We explained that this may be because education promotes the value of community service and civic participation. In addition, we speculated that better educated older people may be more likely to participate in formal organizations because they develop "bridging social capital" while in school which results in wider social networks that draw them into participation. More highly educated older people also have more financial resources and can afford to pay dues and participation fees. Morevoer, they can better afford transportation and other costs associated with community involvement.[2] While factors other than education show the expected directions of relationship with organizational participation among in-migrants, none reached the level of statistical significance. Among non-migrants, on the other hand, participation was positively associated with duration of residence and better health and negatively associated with having adult children living within a half hour drive.

We have included education as a predictor of changing health status in the analysis of this issue presented later in the chapter. We argue that more highly educated persons will be healthier than persons with less education partly because they are more likely to participate in organizations, which has a positive impact on health. However, since education and organizational participation share an association with health status, one or the other or both of their independent effects could be depressed in a multivariate analysis that includes both factors simultaneously.

IN-MIGRANTS' AND NON-MIGRANTS' SOCIAL PARTICIPATION
AND HEALTH

Social Integration

We begin our analysis by describing levels of social integration and
health within both the in-migrant and non-migrant samples, comparing data
from the 2002 and 2005 waves of the survey. We initially hypothesized that in-
migrants would be substantially less integrated in their destination communities
than non-migrants, given that in-migrants in 2002 had lived in a RRD on average
2.9 years and non-migrants on average 22.1 years. However, an earlier analysis
of the 2002 data (Glasgow and Brown, 2006) showed that, while in-migrants
had fewer primary group relationships in destination communities than non-
migrants, a surprisingly high proportion of newcomers had already developed
various types of informal social relationships. Moreover, the data showed that
they were participating in formal organizations and activities at a relatively high
level compared with longer-term older residents (see Figure 5.1).

*Figure 5.1. Older In-migrant Volunteer with the Director of Suttons Bay,
Michigan's Senior Center*

Table 5.1 displays indicators of the informal social networks of older in-migrants and non-migrants in 2002 and 2005. As expected, a larger proportion of the non-migrants than migrants have primary group relationships. In 2002 and 2005 half of non-migrants had at least one child; about 40 percent had at least one grandchild; and over 40 percent had other relatives living nearby. Nevertheless, though lower, fully a third of in-migrants had at least one child within a half hour drive, almost a third at least one grandchild, and a third had other relatives living nearby in both years. Moreover, the migrants and non-migrants average the same frequency of visits with friends (1–2 times per week) in both 2002 and 2005 and there is very little difference between migrants and non-migrants in the percentages seeing their children at least five times per year. The comparison across the two groups at two different points in time show remarkable stability in the level of participation in informal social networks between the two waves of the survey. The changing composition of the two samples between the two waves of the survey could account for some of the similarity in responses between 2002 and 2005, if relatively less healthy individuals were those who dropped out of the survey between the first and second wave of data collection. It could also

Table 5.1. Informal Social Relationships of Migrant vs. Non-Migrant Older Residents of Rural Retirement Destination Counties, 2002 and 2005

	2002		2005	
	Migrants (N = 368)	Non-Migrants (N = 420)	Migrants (N = 282)	Non-Migrants (N = 321)
% with at least 1 child within ¹/₂ hr drive	34.9	49.7	33.6	52.6
% with 2 or more children within ¹/₂ hr drive	8.4	24.2	8.5	27.5
% with at least 1 grandchild within ¹/₂ hr drive	28.6	39.7	30.7	43.2
% who see their children 5 or more times/year	78.8	86.5	75.8	82.4
% with other relatives within 1 hr drive	31.8	42.0	33.8	47.0
Average frequency of visits with friends	1–2 times/wk	1–2 times/wk	1–2 times/wk	1–2 times/wk

Source: Cornell Retirement Migration Survey.

be that the approximately three-year time lag between the survey dates was not long enough for very much actual change to have taken place.

Formal social participation of migrants and non-migrants is shown in Table 5.2. The 2002 data show that modest differences exist between the two groups with respect to particular types of participation, whether it be through service and political organizations or volunteering and religious services attendance. Longer-term residents had a somewhat higher level of participation than the in-migrants on each of the four types of formal social participation, but, except for participation in service organizations, the differences while systematic are not especially large. In addition, migrants and non-migrants were about equally likely to participate in social clubs and the senior center (see Figure 5.1). These findings are striking, given the much shorter duration of residence of the migrants than the non-migrants. These data indicate that older in-migrants to RRDs are succeeding in their efforts to become socially integrated in their new communities.

Even more remarkable, by 2005 the in-migrants had become *more* likely than non-migrants to participate in service, social and volunteer organizations/activities (see Figures 5.2 and 5.3), while non-migrants had a slight edge over the in-migrants in the proportions participating in political, religious and

Table 5.2. Formal Social Relationships among Migrant vs. Non-Migrant Older Residents of Rural Retirement Destination Counties, 2002 and 2005

	2002		2005	
	Migrants (N = 368)	Non-Migrants (N = 420)	Migrants (N = 282)	Non-Migrants (N = 321)
	Percents			
Participate in service organizations	22.8	31.0	44.3	42.5
Participate in political organizations	8.9	13.1	18.4	20.6
Participate in social clubs	32.3	32.6	35.5	32.2
Participate in volunteer activity	38.1	42.6	47.5	42.4
Never attend religious services	36.3	32.1	35.2	34.3
Attend senior center, *sometimes or often*	12.3	12.6	12.0	12.2

Source: Cornell Retirement Migration Survey.

*Figure 5.2. In-migrant Returned Home to Give Something Back
to His Community*

senior center activities (Table 5.2). As noted earlier, some attrition occurred
in the migrant and non-migrant samples between 2002 and 2005. Could it be
that sample attrition influenced findings observed for 2002 compared to 2005?
It is possible, but given that attrition occurred in both the in-migrant and the
non-migrant samples, that is probably not the explanation. More likely, the in-
migrants exhibit a comparatively strong tendency to become involved in their
new communities, and RRDs seem to provide ample opportunities for them
to do so. In addition, previous research showed that in-migrants to RRDs are
positively selected on such characteristics as income, education and age (see
Chapter 4)—factors shown to predict higher levels of formal social participation
(Glasgow and Brown, 2006). Becoming involved in RRD communities is one
way that older newcomers adapt to their change in residence.

   Also worthy of note is that, of the various types of participation, it was
only with regard to attending religious services that more than fifty percent of
both in-migrants and non-migrants participated at some level. Accordingly, even
though a substantial proportion of older persons are socially active in RRDs,
there is room for improvement. With a large body of past research showing
that social participation is positively associated with health and longevity

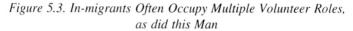

*Figure 5.3. In-migrants Often Occupy Multiple Volunteer Roles,
as did this Man*

(Moen, Dempster-McClain and Williams, 1989), RRD officials and community
leaders might do well to encourage and provide opportunities for further formal
participation of older newcomers and longer-term residents alike. However, our
face-to-face interviews with a small number of in-migrants in four case study
sites revealed that older in-migrants perceived few barriers to participation in
community organizations and activities, so we are not sure how community
leaders could promote an even higher level of involvement.

Health

Descriptive analysis of health and functional ability indicates that the
in-migrants are in somewhat better health than the non-migrants, a situation
that holds in both 2002 and 2005 (Table 5.3). But the differences between the
two groups are modest and thus surprising, given that the non-migrants are
older on average than the in-migrants (see Chapter 4). Between 2002 and 2005
self-perceived health declined slightly among both migrants and non-migrants;
but the two groups were less likely to report the occurrence of an illness
or injury within two years of the last interview. Depending on the particular

Table 5.3. Health Among Migrant vs. Non-Migrant Older Residents of Rural Retirement Destination Counties, 2002 and 2005

	2002		2005	
	Migrants (N = 368)	Non-Migrants (N = 420)	Migrants (N = 282)	Non-Migrants (N = 327)
	Percents			
Health Status:				
Rating health good or excellent compared to others	85.8	81.9	84.6	79.3
Having illness or injury during past 2 yrs	28.3	31.4	26.7	28.8
Activity Limitation with respect to:				
Walking 6 blocks	22.0	29.0	24.4	26.5
Climbing stairs	19.0	21.1	19.0	21.3
Doing household tasks	12.8	16.0	12.9	17.3
Going shopping	8.1	10.2	7.3	10.0
Volunteering	9.1	16.4	10.6	14.4
Driving a car	8.1	9.1	7.3	8.4
Participating in recreation	18.8	25.5	20.4	25.0
Bending, kneeling, stooping	26.9	26.3	25.8	29.9

Source: Cornell Retirement Migration Survey.

measures of functional limitations, in-migrants and non-migrants showed both increases and declines over time. These findings may reflect fluidity in having problems that limit one's activities of daily living, but these findings may also be due to the sample attrition that occurred between the two waves of the survey.

With samples of older people, one would expect worsening health indicators over time. But it should be stressed that around 85 percent of in-migrants and approximately 80 percent of non-migrants in both 2002 and 2005 rated their overall health as good or excellent, and less than a third of both groups was diagnosed with a new illness within two years prior to each interview. The overall conclusion to be drawn from these findings is that older residents of the 14 RRD counties, both in-migrants and non-migrants alike, are in remarkably good health.[3]

WHAT FACTORS ACCOUNT FOR DIFFERENCES IN HEALTH
AND DISABILITY AMONG OLDER PERSONS LIVING IN RRDs?

Our primary focus in this analysis is to determine whether social integration is associated with variability in health status among older residents in RRDs. We hypothesize that participation in formal organizations will have a stronger effect on health status than will involvement in informal social relationships. This broad encompassing definition and measurement of social integration makes it important to distinguish between *informal* participation in a network of social ties and *formal* participation in civic and cultural organizations in the community (Young and Glasgow, 1998). As noted earlier, most previous research has focused primarily on the effect of the social support individuals receive from informal social relationships on health status. Here, we provide a competitive test of whether formal or informal social participation is more important in predicting the health and disability status of older persons living in 14 RRDs. We compare the impact of social integration on the health status of migrants versus non-migrants, and we use weighted data to examine these relationships for the overall sample where migration status is entered into the equation as a control variable. As indicated in Chapter 1, migration among older persons is relatively rare. Accordingly, we over-sampled in-migrants in order to obtain sufficient numbers for analysis. In order to pool in-migrant and non-migrant data into a representative sample of older persons in the 14 RRDs, it is necessary to weight the data by taking into consideration the actual representation of in-migrants and non-migrants in the older populations of RRDs (see Chapter 4 for an explanation of the weighting procedure used in analyses of the overall sample, which merged data for in-migrants and non-migrants into one pooled sample). We controlled for migration status because, as discussed earlier, the act of migration tends to disrupt older persons' social relationships. However, though that may be true, the descriptive data displayed in Tables 5.1 and 5.2 show that many in-migrants have joined family and friends in RRDs and have succeeded in establishing organizational and community ties in their new destinations. Accordingly, while we still expect migration status to have a depressive effect on social integration, and therefore on health and disability, we do not expect this effect to be especially strong.

We have only two-waves of data in our panel survey, which is not ideal for conducting a longitudinal analysis of predictors of change in health status (or any other topic) (Singer and Willett, 2003), but we were unable to secure funding for additional waves of data collection.[4] On the plus side, these data are unique among data sets on retirement migration or social integration, and we believe much is to be learned from our analysis. Nevertheless, future studies would do well to conduct longitudinal studies over a longer time horizon and include more waves of data collection than our study offers.

Variables and Measurement

Dependent Variables

Before turning to the results, we describe the dependent and independent variables used in the analysis and discuss how each is measured. We use two types of health measures, each tapping a different dimension of health status, as dependent variables in separate OLS regression equations. These two variables are functional limitations and self-perceived health, both measured at time 2. The index of functional limitations was created based on respondents' reports on whether they had difficulty performing eight physical activities, including walking six blocks, climbing stairs, doing household tasks, going shopping, volunteering, driving a car, participating in recreation, and bending, kneeling or stooping. For each of these functional activities, a score of 0 is assigned for "no difficulty" and 1 indicating that respondents report "some difficulty" performing them. Accordingly, the additive index of functional limitations ranges from 0, indicating no difficulty on any of the activities enumerated, to 8, which implies some difficulty in performing all of the aforementioned functions.

The self-rated health dependent variable reflects how respondents perceived their overall health at time 2 by answering a single 4-point scale question ranging from "very poor" to "excellent." While this measure is subjective, past research has demonstrated a significant relationship between self-rated health and mortality. Idler and Kasl (1991) conclude that, net of controls, the statistical relationship between self-rated health and mortality is valid and that perception of poor health is a cause of early mortality. Thus, we believe a measure of self-rated health has validity for our analysis as well.

Independent Variables

As indicated above, our principal motivation in conducting this analysis is to examine the associations between formal and informal social participation and health. Our measures of informal social participation include the number of children living within a half hour drive of the respondent, the number of other relatives within 1 hour's drive (excluding those living in the respondent's household and adult children), and an additive scale of the frequency of visitation with relatives, friends and neighbors, with values ranging from 3–22. Measures of formal participation include frequency of religious services attendance (1–7 range in response categories), number of types of organizations belonged to, including social, service, political and volunteer organizations (0–4 range), and frequency of participation in a senior center (1–7 range). In our analysis, the measure of religious services participation is kept separate from the additive index of the number of types of organizational participation because religious affiliations are sometimes ascribed from birth, and thus religious participation may not be altogether voluntary (Young and Glasgow, 1998). Our general expectation is that, when other factors are controlled, these measures of formal social participation will have a marked positive effect on health and a depressive effect on disability.

We have kept participation in a senior center separate as well because Young and Glasgow's (1998) analysis showed an association where more disabled and less healthy individuals were more likely than healthier older people to participate in senior center activities. Young and Glasgow, in a factor analysis, produced findings suggesting that participation in a senior center is a type of "compensatory adaptation" used primarily by persons who have reduced social contact because of deaths or out-migration of friends, relatives and neighbors and/or because they have become less geographically mobile. With regard to this particular type of formal participation, therefore, we expect to find a *negative* association between senior center participation and higher levels of health.

Included among the independent variables are controls measuring socio-demographic and socio-economic status (SES) variables typically included in sociological investigations of the determinants of health. These control variables are: age (measured in years), marital status (0 = not married, 1 = married/partnered), gender (0 = male, 1 = female), employment status (0 = not employed, 1 = employed), years of education (measured in categories ranging from 0–19) and household income (measured in 5 categories, including less than $10,000, $10,000–$25,000, $25,000–$49,999, $50,000–$75,00 and $75,000 and over, with this final category used as the reference category in a dummy coded variable).

In addition, we include two control variables not typically used in analyses of the determinants of health status—frequency of doctors' office visits (measured as the number of times a respondent saw a physician in the past year) and the number of overnight stays in a hospital during the past two years. These items are used as controls because any significant findings for social participation or SES could be challenged on the grounds that they are indirectly measuring use of medical care (Young and Glasgow, 1998).

Change in Health and Activity Limitations, 2002–2005

One's baseline health and disability status are likely to have a strong influence on health and disability status measured at a later time. Accordingly, we included functional limitations at baseline, or time 1, in equations predicting functional limitations at time 2. Similarly, self-perceived health at time 1 is included as a control in equations predicting self-perceived health at time 2. By-doing this we have conducted an analysis of *change in health status and functional limitations* rather than an analysis of variability in health status or activity limitations at time 2. This procedure adjusts for the effect of initial value on change. The higher the initial value of health status or activity limitations, the greater absolute increase in health status or activity limitations between the two waves of the survey (Long, 1979). In other words, it is expected that persons with a health advantage will retain or increase their advantage over time, and vice versa. This procedure is preferable to using the percentage change as a dependent variable since percentage change is negatively correlated with initial value. In other words, the dependent variable is the difference between the observed level of health status or functional limitations at wave 2, and the expected value is based on the partial relationship between waves 1 and 2 across all respondents. Since this is an analysis of change, all of the independent variables used were those measured during wave 1 of the survey.

Findings

Factors Associated with Changes in Activity Limitations

Table 5.4 displays regression coefficients and standard errors for variables used to predict change in functional limitations. This table includes three different models. Model 1 is for all survey respondents living in the 14 RRDs, and it includes a variable indicating whether the respondent is a migrant or not. In models 2 and 3 the analysis is run separately for older in-migrants and non-migrants. In models 2 and 3 the variables are the same except that there is no control for migrant status, and we do not use weighted data for the analyses conducted separately for in-migrants and non-migrants. Among the measures of informal social ties, only the number of other relatives living nearby was significant in its effect on changes in the number of functional limitations. Contrary to our expectations, the relationship was positive and at the .10 statistical level. This indicates that older residents of RRDs with a relatively *greater* number of "other relatives" living nearby reported *a greater increase in* functional limitations than those who had few or no relatives of this type living in close proximity. The statistically significant relationship, however, held only for the pooled sample of all respondents in the 14 RRDs. This finding might seem counterintuitive, but it may be that "other relatives" (those who are not adult children and those not living in a respondent's household) add stress to older individuals' lives, in turn having something of a deleterious effect on their functional ability. Mainly, what these findings show is that, after controlling for other variables, informal social

Table 5.4. Determinants of Change in Functional Limitations among Older Residents of Rural Retirement Destinations, 2002–2005[a]

Independent variables	All respondents (weighted) (N=388)	Migrants (N=274)	Non-Migrants (N=312)
Intercept	−1.469(1.488)	−1.337(1.732)	−1.384(1.663)
Age	.062(.017)***	.040(.019)*	.070(.019)***
Respondent is married/partnered	.347(.237)	−.129(.291)	.489(.262)+
Respondent is female	.248(.222)	.175(.232)	.285(.261)
Education	−.124(.045)**	−.032(.048)	−.154(.052)**
Respondent is employed	−.426(.333)	−.366(.331)	−.414(.391)
Income			
Less than $10K	−.221(.644)	1.168(.745)	−.670(.735)
$10k–$25,000	−.314(.435)	.286(.446)	−.637(.515)
$25,000–49,999	−.380(.368)	.194(.342)	−.646(.446)
$50,000–75,000	−.652(.413)	.613(.387)	−1.061(.492)
$75,000–reference	–	–	–
Freq. of doctors' office visits	.248(.097)*	.261(.109)*	.238(.108)*
Number of overnight stays in hospital	.153(.087)+	.190(.086)*	.125(.103)
Number of children nearby	−.004(.083)	−.237(.143)	−.003(.089)
Number of relatives nearby	.026(.016)+	.025(.016)	.027(.018)
Visit friends, relatives and neighbors	−.036(.029)	−.024(.036)	−.042(.032)
Freq. of religious services attendance	−.091(.041)*	−.032(.047)	−.109(.047)*
Number of types of organizational memberships	−.091(.091)	−.189(.105)+	−.061(.102)
Participation in senior center	.038(.116)	−.179(.128)	.113(.133)
R is a migrant	−.016(.253)	–	–
Number of functional limitations: wave 1	.540(.043)***	.528(.050)***	.547(.049)***
R2	.456	.467	.469
Adjusted R2	.428	.429	.436

Source: Cornell Retirement Migration Study; * p < 0.05; ** p < 0.01; *** p < 0.001; +p < 0.10.
[a] Numbers in the table are regression coefficients, and standard errors (in parentheses).

relationships do not affect change in health status, as measured by functional limitations, of older in-migrants or non-migrants in RRDs.

Measures of formal social participation are somewhat more predictive of functional limitations than the informal measures, but they are still not particularly strong predictors of how older rural residents' ability to function in their daily activities changes over time. For the weighted sample of all respondents and in the model for the non-migrant sample, those who attend religious services less frequently are significantly more likely to report increasing functional impairment (Table 5.4). Why religious participation does not affect the functional limitations of in-migrants is not entirely clear. But it appears that religious participation is more salient to the health status of the longer-term older residents than recent in-migrants. (In Table 5.5, we report a similar finding for the effect of religious participation on the self-reported health of non-migrants versus in-migrants.) Religious participation may not mean as much to the newcomers as it does to longer-term older residents of RRDs because it takes time to establish meaningful roles and relationships through religious participation. In fact, in our face-to-face interviews with selected respondents in four case study RRDs, some older in-migrants told us that they participated in religious congregations largely to gain access to social activities rather than for spiritual reasons.

Similarly, changes in functional limitations are not strongly affected by participation in clubs, organizations and volunteer activities. Only among the in-migrants did the number of different types of organizations that respondents belong to significantly predict change in the level of functional health. This relationship is in the expected negative direction, whereby in-migrants with relatively higher participation in formal organizations are less likely to report functional impairments (Table 5.4). Participation in a senior center was unrelated to changes in functional impairment for the total, in-migrant and non-migrant samples.

The control variables do a better job of predicting change in functional limitations than is true of the measures of social integration. As expected, having functional limitations at time 1 strongly and positively predicts number of functional impairments at time 2 (Table 5.4). Increasing age was positively associated with functional limitations across all three samples—the total, migrant and non-migrant. A relatively higher education was negatively associated with level of functional impairment for the total and the non-migrant samples but not the migrant sample. The frequency with which older rural residents see doctors is significantly related to a higher incidence of functional limitations, with the finding holding across the total, the migrant and the non-migrant samples. The number of overnight stays in a hospital positively and significantly affected number of functional limitations among migrants but not the other two samples (Table 5.4). The relationship is in the expected positive direction for non-migrants, but it fails to reach statistical significance. It is interesting to note that in the pooled sample, migration status did not affect change in functional limitations, indicating that the different findings for migrants and non-migrants are associated with the differing sizes of the two sub-groups in the overall population and their differential impacts on the characteristics of the pooled total, not migration status per se. In other words, being an older in-migrant does not increase or decrease one's prospects of having activity limiting conditions, at least over this short time horizon.

To summarize, neither formal nor informal social participation is a particularly strong predictor of changes in activity limiting conditions. To the extent that participation does affect this measure of health, however, formal social participation tends to have a stronger effect than informal social ties. And the different measures of formal participation do not have the same effects for in-migrants versus non-migrants. Religious participation is a more important predictor for the non-migrants; and participation in a diverse set of clubs, organizations, and community activities is more important to in-migrants' functional health. Neither informal nor formal social participation, however, was as strongly predictive of changes in functional health status as some of the control variables. As will be shown in the case studies reported in Chapter 6, volunteering and becoming actively involved in their new community were important to in-migrants in a wide variety of ways. Thus, it is not surprising that formal participation would positively effect health, as measured by changes in functional ability of in-migrants.

Factors Associated with Changes in Self-Reported Health Status

Table 5.5 presents an analysis of the determinants of change in self-rated health, again comparing across the weighted total, the migrant and the non-migrant samples. Self-rated health is coded in the positive direction, with better health on this measure being higher on the scale. Accordingly, positive coefficients indicate that a particular factor contributes to improving health. The data reported in Table 5.5 indicate that none of the informal social network variables significantly predicts changes in self-perceived health.[5] On the other hand, several of the measures of formal social participation do seem to affect the health of older persons living in RRDs. Parallel to the analysis for changes in functional limitations, frequency of attendance at religious services is positively related to improving self-rated health for the total weighted and the non-migrant samples but not the in-migrant sample. Similarly, we observe a significantly positive association between the number of types of organizations one participates in and changes in self-perceived health among in-migrants, but not for the total weighted and the non-migrant samples (Table 5.5). Participation in a

Table 5.5. Determinants of Change in Self-Rated Health Among Older Residents of Rural Retirement Destination Counties, 2002–2005[a]

Independent variables	All respondents (weighted) (N = 386)	Migrants (N = 273)	Non-Migrants (N = 310)
Intercept	1.161(.513)*	1.634(.714)*	1.082(.559)*
Age	−.001(.006)	.001(.007)	−.002(.007)
Respondent is married/partnered	−.024(.081)	−.030(.111)	−.016(.088)
Respondent is female	.072(.076)	.095(.089)	.061(.088)
Education	.005(.015)	−.002(.018)	.008(.017)
Respondent is employed	.147(.115)	−.001(.127)	.199(.133)
Income			
Less than $10K	−.079(.220)	−.357(.285)	−.012(.247)
$10k–$25,000	.118(.148)	−.050(.171)	.150(.173)
$25,000–49,999	.040(.126)	−.038(.131)	.070(.150)
$50,000–75,000	.179(.141)	.060(.148)	.214(.165)
$75,000–reference	–	–	–
Freq. of doctors' office visits	−.039(.033)	−.118(.041)**	−.017(.037)
Number of overnight stays in hospital	−.110(.030)***	−.030(.033)	−.137(.034)***
Number of children nearby	−.041(.028)	−.053(.055)	−.041(.030)
Number of relatives nearby	.001(.005)	−.004(.006)	.003(.006)
Visit friends, relatives and neighbors	.009(.010)	.018(.014)	.006(.011)
Freq. of religious services attendance	.034(.014)*	.018(.018)	.039(.016)*
Number of types of organizational memberships	.042(.031)	.094(.040)*	.027(.034)
Participation in senior center	−.016(.039)	−.092(.049)+	−.002(.044)
R is a migrant	.072(.089)	–	–
Self-rated health: wave 1	.512(.045)***	.432(.061)***	.533(.049)***
R2	.394	.330	.417
Adjusted R2	.362	.283	.381

Source: Cornell Retirement Migration Study; * p < 0.05; **p < 0.01; ***p < 0.001; +p < 0.10.
[a] Numbers in the table are regression coefficients, and standard errors (in parentheses).

senior center is significant in the in-migrant sample, but it is not statistically associated with changes in health in the total weighted and the non-migrant samples. Moreover, the relationship is negative, suggesting confirmation of a process of compensatory adaptation by participating in a senior center when other types of participation become limited due to poorer health (Young and Glasgow, 1998). It appears that senior centers provide a protective environment of sorts for older individuals whose health is in decline.

Among the control variables, self-rated health in 2002 was strongly predictive of self-rated health in 2005 across all three samples (Table 5.5). This is not surprising, as people who start off in better health do better over time, especially given that age is controlled in the model. A higher frequency of visits to doctors' offices negatively affected changes in self-perceived health among in-migrants but not among the total weighted and the non-migrant samples. The number of overnight stays in the hospital was negatively associated with self-perceived health among the total weighted and the non-migrant samples but not the in-migrant sample. It is not clear why different types of medical services utilization would have different effects for in-migrants versus non-migrants' self-rated health, but such is the case. Similar to the case for changes in functional limitations, being a migrant neither increases nor decreases changes in self-reported health. Once again, this is not surprising given the relative lack of migrant versus non-migrant differences in the comparative profile of health status shown in Table 5.3.

What might seem particularly surprising in this analysis is the relative lack of impact of age, income and other control variables, as well as only a modestly significant impact of participating in formal organizations. It should be noted that some of the variability associated with these factors has been absorbed by including the initial state of the dependent variable as a predictor. In a previous analysis (not shown) where health status at time 1 was not included, age had a negative impact on self-perceived health; doctor visits negatively affected the health of migrants and non-migrants alike; having relatives nearby negatively affected the health of longer-term residents; and, most importantly, organizational participation had a much stronger positive impact on health, regardless of migrant status (Glasgow, 2006).

To summarize these findings, measures of formal social participation have a stronger effect on changes in self-perceived health than do measures of informal social participation, although neither is particularly strong. Informal participation is completely unrelated to changes in health, while participation in formal community organizations seems to have benefits for migrants and church attendance has a positive health impact for longer-term older residents. These findings are not overwhelming, but they do sustain our confidence that older migrants benefit from participating in clubs, organizations and volunteer activities. The analysis also provides further evidence that informal and formal social participation are distinctly different and that formal participation is a stronger predictor of health status than informal social participation. These findings combined with the findings from our case studies (see Chapter 6) show that formal social participation benefits in-migrants' well-being regardless of whether it is in regard to their health status or other aspects of their lives.

CONCLUSIONS

A robust social science literature indicates that older persons benefit from social integration and that successful aging is contingent on living in communities that provide opportunities for engagement in multiple roles, social support and social integration. (Wethington, Moen, Glasgow and Pillemer, 2000). Since migration is typically considered to be a permanent or semi-permanent move of sufficient distance and duration to disrupt everyday social relationships (Long, 1988), we wondered if older persons who moved to RRDs might be at risk of social isolation, and hence of being exposed to the risk of poorer health and other aspects of reduced well-being.

Our analysis showed that these concerns are, by and large, misplaced. Older persons who move to RRDs appear to have little difficulty becoming socially integrated. Moreover, in-migrants appear to have a rich store of both informal ties and formal links to organizations, clubs and volunteer activities. We were surprised to learn that more than one third of in-migrants had at least one adult child living within a half hour drive and that many had other relatives nearby as well. It appears, then, that older persons move to rural areas to take advantage of the scenery and rural ambiance, but they also come to be close to their children, grandchildren, and perhaps even aging parents. Similarly, we found evidence in our survey and in case study interviews with a select group of in-migrants that they had either moved to a RRD to join adult children or that adult children had moved to be nearer to older in-migrants. In other words, their RRD in-migration appears to combine both "amenity- and assistance-driven" motives (Litwak & Longino, 1987). This raises the question of whether older in-migrants with family close by will be less likely to return to their origin community later in life, compared with in-movers who lack similar social relationships. As will be elaborated more fully in Chapter 7, this aging-in-place versus onward migration scenario has important implications for both individual- and community-level aging.

We were also surprised with the high degree of formal social partic-ipation reported by in-migrants to RRDs. Even though they had only lived in these communities for an average of about three years when first interviewed, they were almost equally as likely to participate in clubs, organizations, volunteer activities and religious congregations as similar aged persons who had lived there on average for over twenty years. Our face-to-face interviews with community

Figure 5.4. In-migrant Recently Elected Mayor of Payson, Arizona

leaders and a small subset of survey respondents in four case study communities indicated that older in-migrants face few impediments to their participation and that local organizations depend on in-movers for leadership, technical expertise and daily maintenance activities (see Figure 5.4).

    In examining the association between social participation and well-being among older residents of RRDs, we hypothesized that formal participation would be more beneficial than close ties with children, other relatives and friends. Our analysis found modest support for this proposition. While formal participation was shown to contribute to improved health and lower levels of activity limitation, informal relationships were either not related to these health outcomes, or they diminished well-being. These findings, while generally consistent with previous research (Young and Glasgow, 1998), are weaker than anticipated. This is partly because our dependent variables were measures of change in health, and little change can be expected over the three years separating the two waves of our panel survey. It is likely that we would have found stronger support for the positive impacts of social integration if the interval between measurements had been wider or if we had had data for more than two time points. In fact, cross sectional research examining variability of these same measures of health among older residents of RRDs in 2005 indicates that this is the case. But even that cross sectional analysis of the positive effects of formal social participation on health status was weaker than expected. Our analysis might be plagued by measurement error, or it is possible that social participation is less effective in promoting health in RRDs than in some other environments.[6]

While the analysis in this chapter produced only modest evidence that social participation contributes to improved health among older in-migrants to RRD counties, and that formal participation has a greater positive impact than informal social ties, our case study research reported in Chapter 6 provides a more nuanced view of social participation among older persons who move to the countryside. Our interviews with community leaders and in-migrants show that older people participate for a wide variety of reasons and that the benefits they derive for this participation are complex, multi-faceted and not easily captured by summary measures like self-reported health.

NOTES

1. We conducted an exit interview with the 35 wave 1 respondents who out-migrated from a RRD by 2005, asking them where they moved to and their reason for moving from the county in which they lived in 2002. We analyzed the social integration and health at wave 1 of the out-migrants, which we separated by their original in-migrant or non-migrant status. We do not show this analysis, but we refer to it briefly in conjunction with our analysis of social participation and health of the migrants and non-migrants who completed interviews in 2002 and 2005.
2. We lacked a measure of income in the 2006 analysis, but we were able to code an income variable from the 2002 data for the analysis used in this chapter.
3. An analysis of 2002 health indicators of the 35 respondents who had out-migrated from a study county by 2005 (data not shown here) revealed that the out-migrants (including 19 originally classified as in-migrants and 16 originally classified as non-migrants) were healthier and reported fewer activity limitations than the overall samples of migrants and non-migrants. We are not sure what this means in terms of motivations for out-migration from a RRD, but poor health and a health-related need to be near an adult child/other informal caregiver or better medical services probably were not the major reason for their out-movement.
4. While our analysis would have been stronger if we had had more than two years of data, OLS regression can be used appropriately with only two waves if the residuals of variables are normally distributed. We evaluated this and were satisfied that OLS was appropriate for examining the present questions.
5. While not statistically significant, it is important to note that having children close by appears to have a slight negative effect on older persons' health status.
6. It is also possible that we are looking in the wrong causal direction and that health predicts social participation and not vice versa. Further analysis using multiple stage least squares might unravel the direction of causation issue. In this study, however,

our model clearly specified that we expected participation to result in changes in health since all predictor variables were measured at time 1, the outcome variable was measured at time 2, and the baseline level of the dependent variable was included as a control.

REFERENCES

Berkman, L.F. (1983). The assessment of social networks and social support in the elderly. *Journal of the American Geriatrics Society, 31*(12), 743–749.

Brown, D.L., and Glasgow, N.L. (1991). Capacity building and rural government adaptation to population change. In C.B. Flora and J.A. Christenson (Eds.), *Rural policies for the 1990s* (pp.194–208). Boulder, CO: Westview Press.

Durkheim, E. (1897). *Suicide: A study in sociology*. J.A. Spaulding and G. Simpson (1951 Trans.). New York: Free Press.

Glasgow, N. (2006 May). *Retirement migration, social integration and health*. Paper presented at the Society, Space and Practice Seminar, University of Newcastle upon Tyne.

Glasgow, N., and Brown, D.L. (2006). Social integration among older in-migrants in nonmetropolitan retirement destination counties: Establishing new ties. In W. Kandel and D.L. Brown (Eds.), *Population change and rural society* (pp.177–196). Dordrecht: Springer.

Idler, E.L., and Kasl, S. (1991). Health perceptions and survival: Do global evaluations of health status really predict mortality? *Journal of Gerontology, 46*(2), S55–S65.

Litwak, E., and Longino, C. (1987). Migration patterns among the elderly: A developmental perspective. *Gerontologist, 27*(3), 266–272.

Long, L. (1988). *Migration and residential mobility in the United States*. New York: Russell Sage Foundation.

Long, S. (1979). The continuing debate over the use of ratio variables: Facts and fiction. In K. Schuessler (Ed.), *Sociological methodology, 1980* (pp. 37–67). San Francisco: Jossey Bass.

Moen, P., Dempster-McClain, D., and Williams, R. (1989). Social integration and longevity: An event history analysis of women's roles and resilience. *American Sociological Review, 54*(4), 635–648.

Moen, P., Dempster-McClain, D., and Williams, R. (1992). Successful aging: A life course perspective on women's roles and health. *American Journal of Sociology, 97*(6), 1612–1638.

Moen, P., Fields, V., Quick, H., Hoffsmeister, H. (2000). A Life Course Approach to Retirement and Social Integration. In K. Pillemer, P. Moen, E. Wethington, and N. Glasgow (Eds.), Social Integration in the Second half of life (pp. 75–107). Baltimore: Johns Hopkins University Press.

Pillemer, K., Moen, P., Wethington, E., and Glasgow, N. (2000). *Social integration in the second half of life*. Baltimore: Johns Hopkins University Press.

Singer, J.D., and Willett, J.B. (2003). *Applied longitudinal data analysis: Modeling change and event occurrence*. New York: Oxford University Press.

Wethington, E., Moen, P., Glasgow, N., and Pillemer, K. (2000). Multiple roles, social integration, and health. In K. Pillemer, P. Moen, E. Wethington, and N. Glasgow (Eds.), *Social integration in the second half of life* (pp. 48–71). Baltimore: Johns Hopkins University Press.

Young, F., and Glasgow, N. (1998). Voluntary social participation and health. *Research on Aging, 20*(3), 339–362.

CHAPTER 6

COMMUNITY CHALLENGES
AND OPPORTUNITIES ASSOCIATED
WITH RETIREMENT IN-MIGRATION

INTRODUCTION

This chapter examines the impacts of in-migration by older persons on the social and economic organization of rural retirement destinations. The chapter's overall focus is to examine community-level opportunities and challenges associated with the in-movement of older persons. It uses case study research to elaborate a number of themes that were introduced in Chapter 2. In that chapter we indicated that our research was largely concerned with *unplanned rural retirement destinations* (RRDs), rather than with planned elder retirement complexes. We explained that, while our study is primarily concerned with the experience of older in-migrants themselves, one cannot understand how older in-migrants establish new social relationships without reference to the social and economic contexts in which this adaptation takes place. Accordingly, in Chapters 2 and 3 we used census data to examine how older in-migration affects the demographic and socio-economic compositions of rural retirement destinations and we discussed how older in-migration appeared to be affecting opportunities available in local economies. In the present chapter we use case study data collected in four of the fourteen counties where our panel survey was conducted in order to dig deeper into these community-level issues. We examine specific ways in which rural retirement destinations are affected by older in-migration and how communities respond to the challenges and opportunities posed by these migration-generated impacts.

In particular, we consider the following questions:

- What aspects of community life are most directly affected by the in-movement of older persons? What institutional adjustments have rural retirement destinations made to accommodate older in-migrants?

- What contributions do in-migrants make to the local economy and society? What costs are associated with the increased number of older persons in rural retirement destinations?
- What is the nature of social relationships between older in-movers and longer-term residents in rural retirement destinations?

A Framework for Examining the Effects of Population Change on Rural Communities

"The age composition of a population shapes community needs and demands for goods, services and economic opportunities, as well as patterns of consumption, life style and social behavior." (Fuguitt, Brown and Beale, 1989, p. 105)

As indicated by the quotation above, the age composition of a community imposes requirements and limitations on each of its institutions. Hence it is through changes in age composition that demand shifts associated with population growth or decline are most clearly articulated. While demographic change affects all institutional domains, its impact is most directly experienced in the local public sector. This is because local government is the institutional nexus through which services are produced and delivered, and plans for future development are prepared and executed (Brown and Glasgow, 1991). During the last two decades, however, local government's institutional centrality with respect to service delivery has declined. Devolution and privatization have altered the organization of local service provision, with the private and not-for-profit sectors assuming greater service providing responsibilities (Warner, 2003). Moreover, as indicated above, population change has a general impact on community organization that far exceeds its effects on service provision. Population change, and in particular changes in age composition, affect the civic sphere, local organizations, local government and the private market. Accordingly, while the case studies reported in this chapter examine how older in-migration affects the local public sector, they also engage population-induced effects experienced in the local economy and by private not-for-profit organizations. While our focus is largely on the impact of older in-migration on community-level processes and structures with specific relevance to older residents, we are also interested in how the in-movement of older persons affects the business climate, the volunteer sector, and other institutional realms that are not necessarily focused on services and opportunities for older persons.

THE CASE STUDY RESEARCH PROCESS

As stated in Chapter 2, our primary data collection strategy in this research was a two wave panel survey of older in-movers and longer-term older residents in 14 purposefully selected rural retirement destinations throughout the U.S. These data provided the primary basis for Chapter 4, where we examined the selectivity of migration among older persons and reasons for leaving their origin communities and selecting a particular destination. In Chapter 5 we also used these data to study the processes by which older in-movers become socially integrated in their new communities, as well as the association between migrants' social integration and their subsequent well-being. These household-level survey data tell us a lot about the migrants themselves, but they reveal nothing about the social contexts in which older migrants are embedded and how these contexts facilitate or constrain their social involvement. Community case studies are an appropriate method for examining these questions. While the preferred strategy would have been to conduct a separate case study in each of the 14 survey counties, we lacked sufficient funding and time to do this. However, we continued to believe that it was important for the case studies to be conducted in a diverse set of locations to gain insights into the wide range of social and economic circumstances in which retirement migration is occurring. We conducted our case studies in 4 of the 14 counties all with markedly different histories and contemporary situations (see Figure 6.1). From the 14 survey counties, we selected one case study location in each of the nation's major geographic regions where retirement migration is taking place. Thus, we selected Lincoln County, Maine (Northeast), Leelanau County, Michigan (upper Midwest), Transylvania County, North Carolina (South), and Gila County, Arizona (Southwest).

Field Research Design

We spent approximately one week in each of the four counties. In each county our study design involved: (a) a series of 10–12 interviews with public officials, business owners, service providers, and organizational leaders, and (b) face-to-face interviews with 6–7 of the older in-movers who had previously responded to both waves of our telephone survey. Face-to-face interviews with survey respondents gave us an opportunity to probe more deeply than was possible with the survey regarding issues such as motivations for community participation, the nature of respondents' social involvements in the community, reasons for moving to the community and concerns and plans for the future.

Figure 6.1. Counties Where Community Case Studies Were Conducted, 2006

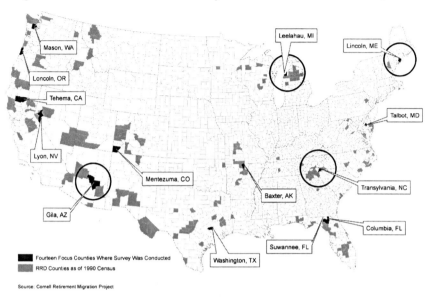

Selecting Interviewees

Each of the four case study counties had a Web site that provided strategic information for setting up the field visits. We studied each Web site to obtain an understanding of how basic functions are provided within each study county. As Warren (1978) has indicated, communities perform both "task" functions, those that mediate between the community and its environment, and "maintenance" functions, those that are internal to the community. Our primary interest in these case studies was to gain an understanding of how maintenance functions are conducted in rural retirement destinations. In particular, we wanted to know how local government is organized in the respective counties, which institutions are responsible for delivering essential services (to older residents and to the community at large), the role played by non-profit organizations in social welfare and service provision, and which non-profit organizations are active in each county.

As indicated above, local government plays a key role in rural communities, so each case study began by interviewing elected and appointed public officials. However, as local government is organized differently in different parts of the U.S., we interviewed different officials in various locations. In Lincoln, Maine and in the rest of New England, for example, towns are the principal unit of local government, so interviews with selectmen[1] were crucial. In contrast, in

Leelanau County, Michigan and elsewhere in the Midwest, counties have more powers and duties, and interviews with county officials revealed more about community response to older in-migration. Once we understood how the public sector was organized, and the roles played by non-profit organizations, we were able to determine which officials and leaders should be interviewed. A visit to the appropriate Web pages helped us to identify the various offices, the identity of current office holders and their contact information. We arranged the majority of these interviews by telephone about one week prior to visiting each county. In general local officials, as well as business and organizational leaders, were enthusiastic about our study and willingly agreed to be interviewed. We also informed the local police and sheriff departments about our study so they would be aware of our activities in their various jurisdictions. Moreover, we contacted local newspapers in each location and asked them to run a story about our study prior to our arrival (see Figure 6.2). This gave us added credibility and helped us gain the cooperation of prospective interviewees whom we met during the course of our field work.

It was not possible to anticipate all of the issues we would encounter in each location prior to our visits, and hence we were unable to pre-schedule all of the interviews before arriving in the respective communities. For example,

Figure 6.2. Editor of Transylvania Times

during our first case study in Lincoln County, Maine, it became evident that
older in-movers were very active community volunteers, but that much of this
activity was not prominently displayed on the community's Web site or in other
promotional materials we had access to when we planned our visits. It was
necessary to arrange some interviews during our visits, typically in response to
recommendations made by local officials, organizational leaders or other key
informants. We also held informal interviews with business owners, employees
and other persons in stores, offices, restaurants and on the street. We did
these as one gauge of the impact of older in-migration on the local business
climate.

We scheduled interviews with public officials early in each visit
believing that local planners, selectmen, county administrators, town supervisors,
mayors and other officials would have a broad view of local issues and events
(see Figure 6.3). We also scheduled visits with local newspaper editors on the
first or second day for the same reason, but also because we felt that journalists
would have fewer political constraints on their views and opinions. During the
middle of our visits we scheduled meetings with aging services providers such as

*Figure 6.3. Director of Planning & Community Development, Leelanau
County, Michigan*

senior centers, Retired Senior Volunteer Programs (RSVP), health care facilities and senior housing units. We also arranged for meetings with chambers of commerce and local business leaders during the middle portion of our visits.

Most of our interviews with older in-migrants who had previously responded to the telephone panel survey were scheduled at the end of each community visit. We separated in-migrants who had responded to both the 2002 and 2005 telephone surveys into persons who were active participators and those with lower levels of social participation. We scheduled about six to seven interviews with previous survey respondents in each study county and we tried to divide these interviews equally between active and less active participants. In reality, however, active participants were more numerous and more likely to agree to a face-to-face interview. For the most part, these interviews were scheduled by phone before we arrived in the community. We reconfirmed appointments by phone after we arrived and only one pre-arranged interview failed to materialize. While the telephone survey selected one older person at random in each survey household, we interviewed couples together during the community visits when they requested that we do so. We feel that this provided some interesting and important insights into factors such as family dynamics, their styles of sociability and their future plans.

The Interview

Interviews with officials and leaders were conducted in their offices and places of business, while survey respondents were mostly interviewed in their homes (see Figures 6.4 and 6.5). Interviews took approximately one hour each. We prepared three different guides to shape the interviews: one for officials and leaders, one for service providers and one for survey respondents. All questions were open ended. Even though we prepared interview guides, interviews were free flowing conversations and did not necessarily follow the question order prescribed by the guide. In these instances, we simply moved back and forth among questions noting information where appropriate on the form. In other words, the guides helped us to organize our questioning, assured us that each respondent would be asked similar questions and provided a structure for organizing responses. Our technique is similar to the ethno-survey used by Douglas Massey and his colleagues in the Mexican Migration Survey (Massey, Alarcon, Durand and Gonzalez, 1987).

Each interview guide focused on a particular set of issues that we wanted to discuss with the various respondents.[2] The "officials" guide began by asking the respondent to share general impressions of how the in-movement of older persons might be affecting the community. This was followed by a

Figure 6.4. Director of Community Development, Payson Arizona

series of questions focused on how the in-movement of older persons might be affecting various aspects of community life, and whether any community decisions or actions had been made that were linked to older in-migration. Then we asked whether they felt that older in-movers "fit" into the community and whether in-movement of older persons had generated any conflict or contention. We next asked whether they felt in-movers were a resource to the community and had them describe the benefits and costs that are associated with having older persons move to the area. We concluded the interview by asking if the community had conducted any programs that were explicitly aimed at recruiting older persons.

Service providers were asked to share their impressions of how the in-movement of older persons was affecting the community and their particular organization (see Figure 6.6). They were then asked about their organization: the services it provided, where in the community it is most active and who its clients are. This was followed by a series of questions focusing on their organization's use of volunteers: what volunteers do, who they are and how they are recruited. The interview concluded by asking respondents whether they felt that their organizations were spaces where in-movers and long time residents come together.

Figure 6.5. Selectman, Southport, Maine

In our interviews with older in-migrant survey respondents, we broke the ice by asking them to tell us something about themselves and, in particular, how long they had lived in the community, where they moved from, how they learned about the destination community and why they chose to move to the area. We then asked them to tell us what they liked and disliked about living in the community. This was followed by questions about the nature of their social relations in the community and whether their friends were mostly in-movers like themselves, had similar socio-economic status and lived near them. Next we asked them a series of questions about their organizational participation including their involvement in formal religious life in the community. We concluded the interview by asking them what features of the community facilitated their organizational participation and whether there were factors that constrained such involvement.

THE FOUR STUDY SITES

In the Northeast, we selected Lincoln County, Maine, a long time "summer place" which has now been transformed into a year-round residential community. Leelanau County, Michigan was our choice in the upper Midwest.

Figure 6.6. Staff of Gila Senior Services, Globe, Arizona

This county is also a summer place of long-standing, but it has yet to retain its summer residents during the winter months. We selected Transylvania County, North Carolina in the South. While summer camps have been important in this county for some time, its economic base was traditionally dominated by high-wage durable manufacturing. Manufacturing declined dramatically and precipitously in the early 2000s, and has been replaced by tourism and retirement living. Our choice in the Southwest was Gila County, Arizona, where timbering, mining and ranching dominated until recently. The southern portion of the county is still heavily dependent on mining (or the lack thereof), but retirees began streaming into the northern town of Payson during the last decade. Accordingly, this county has the shortest history as a retirement destination of any of our case study sites.

In this section, we use census data to describe and compare the socio-demographic characteristics of the four rural retirement counties where we conducted in-depth case studies. We then profile each of the four counties examining their respective geographic locations, histories, economic activities and the organization of their local governments. These analyses provide a context within which to consider the community-level impacts of older in-migration.

Table 6.1 contains a comparative profile of the four rural retirement destination counties and the same indicators for the nation's 2250

Table 6.1. Comparative Profile of Rural Retirement Destinations and All Nonmetropolitan Counties, 2000

	Total	Gila	Leelanau	Lincoln	Transylvania
Demographic					
Population (000)	24.3	51.3	21.1	33.6	29.3
Density (per					
sq/mi)	21.7	10.8	60.6	73.7	77.5
Population change					
(%)	10.3	27.6	27.8	10.7	14.9
Net migration					
(%)[a]	6.9	6.2	9.5	1.7	9.3
Percent					
Non-Hispanic					
white	83	77.8	93.5	98.5	93.7
Adjacent to metro[b]	–	yes	yes	no	yes
Median Age (yrs.)	37	42.3	42.6	42.6	43.9
Percent 65+	15.5	19.8	17.4	18.2	21.4
Socio-economic					
Median H.H.					
income ($000)	32.3	30.9	47.1	38.7	38.6
Percent Poverty	14.9	12.6	5.4	6.6	6.6
Median house					
value ($000)	72.5	100.1	165.4	120	122.3

Source: U.S. Census of Population, SF3, 2000.
[a] In movers age 5+ as a percent of the mid-decade population.
[b] 44% of all nonmetro counties were adjacent to metro counties in 2000 (Brown, Cromartie and Kulcsar, 2004). Gila is adjacent to Phoenix, Leelanau to Traverse City, and Transylvania to Asheville.

nonmetropolitan counties. The first observation is that while the four case study counties are quite variable, some overall comparisons can be made between them and all nonmetropolitan counties taken together. The four study counties are about average in population size; but, with the exception of Gila, they have substantially higher population densities than the rest of rural America. Three of the four grew considerably faster than other rural counties between 1990 and 2000. Only two of the four case study counties had higher overall net migration rates than the rural total, but it should be remembered that all four had at least 15 percent in-migration at age 60 and above, which is considerably higher than the rural total. They are significantly older than the total rural sector, and three out of four are better off with respect to income, poverty and housing value. Three out of four of the case study counties were adjacent to a metropolitan area, even though the residential concentration of older in-migrants in two of these adjacent counties, Gila and Transylvania, is located at a substantial distance from the nearby metropolitan city. Each of the four counties is now discussed separately so that the interviews with local leaders, officials and decision makers that follow can be contextualized.

Gila County, Arizona

Gila appears to be quite different from the other three case study counties. It has a larger population, is less dense (because of its large land area), less well off and has a much higher minority percentage of the population. Moreover, permanent migration of older persons is a relatively recent phenomenon in Gila compared with many other rural retirement destinations. Many older in-migrants moved north to Gila from Phoenix in search of cooler weather and lower real estate prices, while maintaining easy access to big-city services and adult children living in the Phoenix area. Gila is in the center of the state, about 90 miles northeast of Phoenix. Until recently, this distance effectively separated Gila from the Phoenix area, but the construction of the new "Beeline Highway" (U.S. Highway 87) has significantly increased accessibility to Phoenix.

Gila County, overall, does not look like the "typical" prosperous rural retirement destination because the county encompasses three separate social worlds—one quite well off, one mired in its resource-dependent, underdeveloped past and one dominated by part-year residents. Year-round retirement in-migration is concentrated in the northern part of the county around the town[3] of Payson (see Figure 6.7). This old timber and ranching area only began attracting older in-migrants about 10–15 years ago. By the 2000 census, in contrast to the county as a whole, Payson's profile had become quite similar to that of the three other rural retirement counties we studied. It grew by 62 percent between 1990 and 2000, had a minority population of only 8.7 percent, had a median household income of $33,638 and a median value of owner occupied housing of $134,000.

In contrast to Payson, the southern part of the county, especially the area located near the county seat of Globe, is much poorer and is still strongly linked with its copper mining past. This part of the county also has a large concentration of lower-income Hispanics and American Indians. American Indians have lived in this area since 300 BC (Rim Country Chamber of Commerce, 2006). The earliest group, called Mogollons, moved to the area from what is now New Mexico. At the present time, over 6,600 of the county's 51,000 residents are American Indians. The Tonto Apache tribe is the most prominent group. Tribal members live primarily in the southern part of the county, but

Figure 6.7. Gila County, Arizona (Payson)

132 members live on the 85 acre reservation near Payson.[4] The Tonto's operate the Mazatzal Casino, the county's largest non-government employer, on the reservation.

Roosevelt Reservoir, located in the middle of the county, attracts a large population of "snowbirds" during the winter. Accordingly, the county's winter population is substantially larger than the number of persons who live there in the summer. This seasonal population is easily identified by the presence of RV campers parked in settlements adjacent to the reservoir. In addition to Roosevelt Lake, the county has a number of other notable attractions that draw tourists and retirees to the area. The heavily wooded Mogollon Rim, a 200 mile long escarpment that rises to an elevation of 7500 feet, cuts through the county about 15 miles north of Payson. The Rim is generally considered one of Arizona's major attractions. Tonto Natural Bridge, the largest travertine bridge in the world, is located in that area of the county (Rim Country Chamber of Commerce, 2006).

Local government in the county is organizationally and geographically divided between Globe, the county seat, Payson and the county's other towns. Because of the county's large geographic size and the demographic and socio-economic differences between its northern and southern parts, some county departments have satellite offices in Payson. Since in-migrating retirees are concentrated in the Payson area, we focused many of our interviews there. We interviewed the town's mayor, planner and manager, and the director of the Payson Regional Economic Development Corporation. However, as most county provided aging-related services are located in Globe, our interviews with *county* officials were held there. Many county-level officials observed that it was difficult to develop programs responsive to serving the varying needs of such a socio-economically and ethnically diverse population, especially since class and ethnic sub-populations are geographically separated in the county. The large part-year population of snowbirds further complicates the situation.

Leelenau County, Michigan

Leelanau is located in the northern part of Michigan's lower peninsula. The county itself is a peninsula which is bordered on the west by Lake Michigan, on the east by Grand Traverse Bay and to the southeast by Traverse City, a rapidly expanding metropolitan center. The county's environment was favorable for fruit growing and it is still the nation's largest producer of cherries. Commercial fishing was also an important industry in the past and some fishing continues to this day. Summer tourism has a long history here, and consequently over one third of its housing stock is seasonal (Leelanau County, 2000). This has serious fiscal implications because properties not classified as primary residences are taxed differentially higher in Michigan. For example, non-homestead properties can be levied school taxes to a maximum of 24 mils (meaning 24 dollars due in taxes for every $1,000 in property value); while homestead properties pay no more than 6 mils (Leelanau County, 2000). Many wealthy families have maintained summer residences in the county for generations, but short duration summer tourism linked with the county's lakes and scenic landscape, the Manitou Islands and the Sleeping Bear Dunes National Lakeshore, is also popular. Recently, the southern part of the county has become increasingly integrated with the Traverse City metropolitan region's expanding commuter shed. In fact, Traverse City's municipal boundary extends into Leelanau County.

Local government is comprised of the county government itself and the county's eleven townships and three incorporated villages.[5] Leland, the county seat, is becoming increasingly separated from the county's changing population and economic activities, and there is a strong movement to transfer the county's functions from Leland to Sutton's Bay, the county's largest settlement (see Figure 6.8). The county also has four school districts, and its changing demographics have generated heated discussions about the need for school consolidation. While the county has doubled in population since 1970, its percentage of school-age population declined from 30 percent in 1970 to less than 20 percent in 2000 (Leelanau County, 2003).[6] The county is 94 percent

Figure 6.8. Leelanau, County, Michigan (Suttons Bay)

non-Hispanic white, although it contains the Grand Traverse Band of Ottawa and Chippewa Indian Reservation. The reservation has a population of about 772 persons and operates the Leelanau Sands Casino. We interviewed the county administrator, the director of the county's planning and community development office, the director of the county office on aging and a number of township level officials.

Lincoln County, Maine

Lincoln is located on the Atlantic coast about 40 miles northeast of the state's largest city, Portland. Since more than half of the distance is via U.S. Route 1 which is often congested, the trip can take an hour or more. Once in the county, travel between its 17 towns can take much longer than would be suggested by considering mileage alone. The association between road mileage and distance can be deceiving because of the craggy nature of Maine's Atlantic coast and the fact that many of the county's towns are located on separate islands and peninsulas (see Figure 6.9). In fact, this small county has over 450 miles of coastline and six rivers, all of which make travel in the county a sometimes slow proposition. Most county-wide functions are administered from Wiscasset,

Figure 6.9. Lincoln County, Maine (New Harbor)

the county seat, but as elsewhere in New England, public management and local governance are situated in the various towns. The county received its name as a compliment to Thomas Pownal, colonial governor of Massachusetts from 1757 to 1769 who was born in Lincoln County, England.

Lincoln's economy has always depended on the ocean. Until recently, fishing, lobstering, clamming, eeling, boat building, forestry and agriculture were its primary pursuits, and these activities are deeply embedded in Lincoln's place identity (Hummon, 1990). Contemporary development is still strongly influenced by the coast, but as a place of residence and recreation, not of marine and other natural resource-based pursuits (Lincoln County, 2006). In fact, according to the 2000 census, less that 5 percent of the county's employed population now works in fishing, farming, timbering and mining.

Summer tourism has been an important part of the county's economy and social life for a century or more. Many families from Boston, New York and elsewhere in the Northeast have been vacationing along the Lincoln County coast for generations. Several well-to-do families have owned or rented "cottages" for many years in the county's picturesque villages, which dot its secluded coves and atmospheric rivers. Boothbay Harbor, in particular, has had a high profile as a somewhat rugged outdoors oriented summer playground for the wealthy. Since contemporary development is so strongly related to ocean access, towns that lack such access have fallen behind on most indicators of economic prosperity, housing and property values. U.S. Route 1 runs diagonally through the county from Wiscasset to Waldoboro, separating coastal communities from the rest of the county and bisecting Lincoln into its better and worse off halves. Since retirement in-migration is much more likely to be along the coast than further inland, it is typically concentrated in towns that are south of Route 1. This means that age and socio-economic status have now combined to delineate Lincoln's axis of inequality and its evolving social structure. Given our focus on older in-migration, we focused our case study in the southern part of the county.

Until recently, Lincoln could be characterized as a "summer place" that closed up after the summer second home and tourist seasons. Now, however, it has become a year-round residence for many persons who previously lived there as "summer folks." This transformation has dramatic implications for the county's economy, service provision, health care system, land use and property values. As will be seen later in this chapter, this transformation has both positive and negative implications for the county. Moreover, the valence of these implications depends importantly on the lens through which they are viewed.

Lincoln County is comprised of 18 towns, each of which elects its own selectmen and manages its own affairs. We interviewed both town- and county-level officials and other leaders, but the majority of our interviews were at the town level. Almost all of our interviews were conducted in towns located south of U.S. Route 1.

Transylvania County, North Carolina

Transylvania is called North Carolina's "land of waterfalls." Surrounded by the Blue Ridge Mountains, the county has over 260 waterfalls, hundreds of acres of state and national forests and an active arts and music community (Brink, 2006). Transylvania has all of the ingredients necessary to be a tourist and retirement mecca, but these activities only became the dominant core of its economy during recent years. This is not to say that tourism, and to some extent retirement, have not been important in the past. Rather, high value manufacturing, which heretofore had anchored the economy, precipitously and unexpectedly declined during the early 2000s, paving the way for tourism and retirement to dominate the local economy.

Transylvania is well known as the home of numerous summer camps, with fifteen presently in operation. Families have also been vacationing in the county for many years, enjoying its abundant outdoor recreational opportunities

and pleasant climate. Retirement living also has a history here and several
retirement communities have existed in the county for decades. The best known
is Lake Toxaway, which was originally built in the latter half of the 1800's
as a playground for Andrew Carnegie, Henry Ford, Harvey Firestone and other
famous industrialists. In its current incarnation, Lake Toxaway is an exclusive
golf and gated mountain community. Retirement living is now much more
widely spread throughout the county, including residences in downtown Brevard.
Many locals observe that retirement has replaced manufacturing as the core of
Transylvania's economy because the accelerated movement of older persons
seemed to coincide in time with local industries' demise.

The county's social and economic life revolves around its county seat,
Brevard (see Figure 6.10). This small city of 6,800 persons is home to Brevard
College, a small liberal arts institution with a particular focus on the fine arts,
and Blue Ridge Community College, a more technically-oriented institution.
Brevard is acclaimed for its music scene, which extends from classical and
jazz to bluegrass and mountain music. The city has several performance venues
including the state-of-the-art Paul Porter Center. Brevard hosts a well known
summer music camp called the Brevard Music Center (BMC). Each summer the

Figure 6.10. Transylvania County, North Carolina (Brevard)

BMC hosts more than 80 concerts showcasing its talented students. Brevard's downtown has been revitalized after years of decline. In 1993, Brevard became an official Main Street City under the aegis of the National Trust for Historic Preservation. Downtown Brevard is reclaiming its role as the county's focal point. Not only does it feature a diverse collection of shops, restaurants and street festivals, but new residents, including retirees, are beginning to purchase or rent homes there.

Our interviews with officials, community leaders and local business owners were concentrated around Brevard. We ranged farther a field when interviewing older in-migrant survey respondents. We interviewed the director of the county planning office, the Transylvania County manager, the mayor of Brevard and the director of the city's planning department. We also interviewed officials with North Carolina State University's Cooperative Extension Service and a number of professionals with the Area Agency on Aging (in Asheville) which serves Transylvania County.

IMPACTS OF OLDER IN-MIGRATION ON COMMUNITY ORGANIZATION

We interviewed 64 elected and appointed officials and leaders of business and not-for-profit organizations during 43 face-to-face sessions in the four rural retirement destination counties. The Appendix to this chapter shows a list of positions held by interviewees.[7] As discussed above, these sessions were organized with an interview guide. Hence, even though the interviews were free flowing, they were all structured around the same set of questions. This permitted us to tease out main themes in how local leaders perceive older in-migration is affecting their respective communities.

We began all interviews by asking the respondents to share their general impressions of how the in-movement of older persons might be affecting their communities. Without exception, these leaders and officials volunteered that older in-migration was having an overall positive impact in their communities. However, we also found evidence that older in-migration was resulting in various types of displacement and contention in all four of the communities.

Positive Impacts

Virtually every person we interviewed agreed that older in-migrants were a *community asset*. In fact, some officials and local leaders were extremely enthusiastic using adjectives such as "huge" and "terrific" when character-izing the positive impacts associated with older in-migration. In-migrants were

observed as bringing money, time, talent, experience, connections and profes-
sional/technical expertise into the various communities. When asked which
aspects of their communities were most directly affected by the in-movement of
older persons, respondents identified the real estate market, construction, health
care, the non-profit sector, churches, arts and cultural organizations and the
commercial sector. Most of the impacts were perceived as positive, but as will be
seen later in the chapter, some impacts, such as those on the real estate market,
tend to cut in both positive and negative directions.

Real Estate Market and Construction

The most obvious impacts of older in-migration are increases in house
building, real estate values, and real estate tax revenues. The positive impacts
on house construction and real estate sales, prices, and taxes were observed in
all four locations, but the impact on taxes was more acute in some locations
than in others. For example, in Leelanau, Michigan, where a high proportion
of older in-movers are seasonal residents, the county benefits from the fact that
these residents own property that is taxed at a much higher rate than similar
priced primary residential property.[8] House construction was reported to be
strong in all four counties we visited, and this had a noticeable effect on the
demand for both skilled and unskilled labor and on the sale of building materials.
Much of this construction takes place in new subdivisions like Star Valley
Arizona, which are not necessarily age segregated, but where there is a high
concentration of retirement-age residents. Some of these subdivisions are gated,
but most are not. Some gentrification and rehabilitation of older structures is
occurring in various villages and small cities such as Brevard, North Carolina,
Payson, Arizona, Boothbay Harbor, Maine and Suttons Bay, Michigan. New
house construction favors property with a view. In Lincoln and Leelanau, this
means beach front property overlooking the water. In Gila and Transylvania, new
housing is frequently built in foothills on lots with a mountain view. In addition,
some new construction is occurring in exclusive, gated communities such as Lake
Toxaway and Connestee falls, North Carolina and Northport Village, Michigan.
These exclusive communities tend to center around golf, tennis, fishing and other
outdoor recreational possibilities. They are not retirement communities in the
sense of providing aging-related services.

A minority of older in-migrants live in continuing care retirement
communities where they begin their tenure in fully independent housing units,
and transition to assisted living or to a nursing facility as their need for care
and assistance increases. While this is not an especially prominent housing
choice at the present time, it may increase as older in-migrants age-in-place in
the rural retirement destinations. In Lincoln County, Maine, for example, where

retirement migration is a long-established process, continuing care retirement communities are well-developed. One in particular is associated with the Miles Health Care System. In Lincoln, where retirement in-migration has been occurring for years and tends to result in permanent year-round residence, it is also resulting in a notable increase in aging-related housing and health services. This has both a positive economic impact, including a strong increase in the demand for nurses, physical therapists and other allied health professionals, and an increase in locally available health care for the entire community. In Gila, Arizona, in contrast, where older in-migration is a more recent phenomenon, the Payson Regional Medical Center is growing and providing more age-related specialties, but we did not observe the development of a continuing care retirement community. In Leelanau, Michigan, the county's only hospital closed recently because it could not compete with a major medical facility in nearby Traverse City and because a high proportion of the county's older population is comprised of seasonal residents. Leelanau meets its increased need for medical care during the tourist season with "docs in the boxes," or temporary doctors' offices in the various villages. It is not possible to simply assume that a greater number of older persons will translate into more health care services, assisted living facilities, nursing homes or continuing care retirement communities. These relationships are contingent on many local specificities, including the area's socio-economic status, its age composition, whether older in-migration is a recent or well established phenomenon, whether older in-migrants reside year-round in the community, the local supply of allied health personnel and competition with nearby larger places.

Business and Commercial Sectors

We had anticipated that retirement in-migration would be perceived to have a general positive impact on local economies, but our interviewees tended to be somewhat circumspect about this. While, as indicated above, older in-migration has a clearly positive effect on house building and the demand for associated professions and suppliers, its more general commercial impact is less easy to discern. Only in Lincoln County, Maine and in Transylvania County, North Carolina, did we receive unambiguous reports that the in-migration of older persons is affecting the county's overall business and commercial climates. Local officials and leaders told us that Lincoln County had recently been transformed from a summer place to a year-round community and that most restaurants and shops now remain open 12 months per year. We were also told that the older in-migration had resulted in the location of many new banks and the expansion of others. Since retirement in-migration to Lincoln began in the 1980s, somewhat earlier than in the other three communities we studied, this more general pattern of economic development may represent a developmental process that will unfold over time if the other retirement destinations continue to attract older persons and if more of these persons become permanent year-round residents. In Transylvania we were told that retirement-associated jobs had replaced employment lost when three large manufacturing plants closed. One respondent commented that retirees, and persons who visit them, like to shop locally. She indicated that the downtown shopping district was much healthier now than before. Another respondent told us that downtown Brevard had a high occupancy rate. For example, almost all downtown retail, commercial and office space was currently rented. She observed, however, that these shops and restaurants catered to high-end tourists and retirees and had displaced establishments with more modest-means clientele.

By contrast, in Payson Arizona, even though older in-migration has completely transformed the community's population size and composition, most respondents observed that the expected commercial dividend had yet to materialize. As one elected official reported, "we don't have too many new stores, and restaurants haven't really taken off." Another Payson interviewee commented that "there isn't a single family shoe store here." Moreover, many respondents explained that stiff competition with Phoenix was stifling retail and commercial development in Payson, even though the only transportation link between the two places is by private automobile and takes more than an hour to drive. Economic development has also lagged in Leelanau, Michigan, primarily because it continues to be a seasonal community. One respondent characterized the county as being "carved out in winter." One agri-tourism entrepreneur commented that the county must develop a year-round tourism industry if it wants to retain high quality labor. He observed that this was a particular problem with respect to emergency medical technicians (EMT) and other first responders, as well as in his own industry.

An important, if seasonal, economic impact associated with older in-migration is what might be characterized as "secondary tourism." Part-year residents tend to have lots of visits from their children and grandchildren, and this translates into restaurant meals, gifts and other expenditures. As one respondent in Leelanau observed, "These folks are the most visited persons you will ever meet."

One final economic impact is that older in-movers are not necessarily retired, and some who arrive in this status later re-enter the labor force. One interviewee in Leelanau County, Michigan divided older in-movers to his community into three categories of economic activity: (a) retirees, (b) part-time workers/part-time volunteers, and (c) full-time employed. Part-time employees tend to work in local restaurants and tourist and/or outdoor recreation-oriented businesses. One respondent told us that local businesses prefer to hire older in-movers over local teens because they are more reliable and mature. Of course, this is not necessarily viewed as a positive impact when seen from the teens' perspective. The full-time employed category is comprised of persons who commute to jobs they held prior to moving to the county and persons who either start their own businesses after moving to the RRD county or seek employment in local businesses. While some commuting is via the Internet, hence not requiring regular physical movement, other persons regularly commute long distances to their jobs. The availability of excellent air service nearby in Traverse City, for example, was reported to facilitate long distance commuting by older executives and consultants who live in Leelanau County. Some entrepreneurs have established new economic concerns in the various RRD counties that employ local workers and generate substantial income and tax receipts. In contrast, some other new businesses are more modest and/or home-based.

Volunteerism

Without exception, local leaders and officials told us that older in-movers play important roles in their communities as volunteers. In fact, in-movers were characterized as the mainstays of many local organizations. One respondent commented that "Just about everything we do in this county uses volunteer labor." In addition to their time, in-movers donate money and, perhaps most importantly, professional expertise. One elected official in Maine confided to us that his jurisdiction had not hired any professional or technical services in a decade. All of the legal, accounting, engineering and architectural services that were needed during this time had been donated free of charge by in-movers. A county official in Michigan told us that he actively recruits in-movers for participation on task forces because of their experience and training. For example, because wastewater runoff into Lake Michigan is a critical problem in Leelanau County, a task force on septic design was recently established. Our respondent reported that in-movers provided most of the expertise on this study group including a PhD geologist, two PhD chemical engineers, and two persons who were retired from careers in municipal waste management.

One possible *negative* economic impact of volunteerism should be indicated. While virtually every leader, official and service provider we interviewed gushed with enthusiasm about the contribution made by older persons with professional and technical expertise, it is possible that the availability of such volunteer assistance undermines the demand for professional workers in the community. In fact, it may be a factor contributing to out-movement of more highly educated younger workers. While volunteerism among in-movers may help local governments balance their books, it may be a drag on economic development and the retention of skilled workers.

What Do Volunteers Do in RRD Communities?

Our four community case studies provide an interesting contrast regarding the kinds of activities for which in-movers choose to volunteer. In Maine we observed a status hierarchy of volunteer activity. In-movers were particularly likely to volunteer for arts and cultural organizations and for the hospital and YMCA. Conversely, service providing organizations, like the senior center, had more difficulty filling their volunteer labor needs for activities such as meals-on-wheels delivery. One interviewee in Lincoln, Maine, told us that the typical volunteer spends about 40 hours per week working for the hospital, suggesting a serious commitment on the part of the volunteer and a major contribution to the hospital. The director of volunteers at Miles Hospital reported that the facility has about 200 volunteers across its 26 service areas. She told us that volunteers are more likely to be longer-term older residents than in-movers because "people from away" (PFA) need a couple of years to get situated before beginning to serve. Similarly, in Leelanau, Michigan, we observed that older volunteers were particularly likely to work for advocacy organizations, such as the land conservancy, but less likely to lend assistance at the Friendship Center.

In contrast, older volunteers appeared to be equally likely to participate in all kinds of organizations in Payson, Arizona. For example, the retired senior volunteer program (RSVP) in Gila County has over 200 volunteers in Payson alone working in the hospice program, as school tutors, in the senior center, in the library and in the family advocacy agency. Other interviewees reported that volunteers provide rides for persons without access to a private automobile, deliver meals-on-wheels and work in the senior center's kitchen and thrift shop. And the Payson Regional Medical Center has an active hospital auxiliary. The hospital runs an adult health and wellness program called Senior Circle, which also depends heavily on older volunteers. We speculated that the lack of a status hierarchy among organizations with respect to their appeal as places to volunteer may be because of the overall lower government provision of social services in Arizona compared with the other states we visited. One interviewee who is associated with senior services in Gila County told us that there had been big cuts in funding for senior services recently in Arizona, despite population aging in places like Gila County. She indicated that this has strained rural infrastructure and increased the need for volunteer help.

Arts and Cultural Organizations

Our community case studies indicated that arts and cultural organizations were strong and growing in Lincoln, Maine and Transylvania, North

Carolina, expanding in Leelanau, Michigan, and emerging in Payson, Arizona. Arts and culture enrich the community's social environment, and can also contribute to economic development. In-movers positively affect arts and culture by increasing the demand for such activities and, just as importantly, by providing financial donations and volunteer labor in arts and cultural organizations.

As indicated in Chapter 4, older in-movers to rural retirement destinations are positively selected with respect to socio-economic status and seven out of ten moved in from metropolitan areas. Many in-migrants have lived where arts and cultural organizations are well-developed and most have had the personal resources to participate in such organizations. Both Lincoln, Maine, and Transylvania, North Carolina, have extensive theater and musical traditions that predate the arrival of older in-migrants. Nevertheless, in both instances support for live performance has expanded markedly in response to the arrival of new residents with a preference for these forms of entertainment. Leelanau, Michigan is also seeing an increase in the demand for live performance, but the seasonal nature of older in-migration to this area limits the support for new infrastructure and organizational development. Moreover, the expansion of cultural venues in nearby Traverse City limits the possibilities for Leelanau to develop its own arts and cultural scene. Similarly, Payson has had trouble developing its own cultural life because so much is available in Phoenix. Compounding this situation is the fact that many older residents in Payson have adult children and grandchildren in the Phoenix area, and they combine family visitation with shopping, medical appointments, attendance at sporting events and concerts and trips to museums.

Older in-movers not only support arts and culture through their patronage, but also donate money and provide volunteer labor. Many retirees served as executives in large corporations and they are skilled managers and administrators who are able to take the lead in organizational development and in fund-raising campaigns. Thus, it is unsurprising that many local leaders and officials, as well as older in-migrants themselves, commented that in-migrants are among the prime movers in arts and cultural organizations. In Lincoln, Maine, for example, in-movers have been the driving force behind the development of the just opened Coastal Maine Botanical Garden. This 250 acre facility is the largest of its kind on the east coast and will likely become a major tourist attraction. In-movers and longer-term older residents provided substantial donations and volunteer labor for the facility's construction and are strong supporters of its maintenance and continuing operation. In Transylvania, North Carolina, older volunteers provide essential support for the Brevard Music Center and other venues of the local music scene. The music scene, however, predates the arrival of older in-migrants and is strongly associated with Brevard College. In contrast, art galleries were non-existent in the county prior to retirement migration. Similarly, in-movers were the prime force behind the development of Brevard's impressive new library. One interviewee told us that "the first one million dollars in funds raised for the library came from retirees." The arts are less well-established in Payson, although an impressive new cooperative art gallery has recently opened which shows painting, sculpture and ceramics produced by a large number of older in-migrants themselves.

Many older in-migrants belong to churches and other religious organizations, attend religious services regularly and participate in church-based activities. Some of the older in-movers we interviewed serve as deacons or in other leadership roles in their churches, others participate in church-based service activities and educational programs. One person we interviewed was a professional choir director prior to moving to Lincoln, Maine, and she serves in a similar role on a voluntary basis in her new community. We found that older in-migrants have a wide variety of motivations for their religious affiliation. Some participate for primarily spiritual reasons, but more of them reported to us that church was a place where they felt socially connected.

Concerns and Challenges Posed by Older In-Migration

While virtually all of the leaders and officials we interviewed were enthusiastic about older in-migration to their communities in some ways, most also confided that in-movement of older persons posed challenges and concerns. Most agreed that older in-migrants are a community resource, but some also felt that older in-migration has had negative community impacts. While the majority of interviewees believed that the positive impacts were relatively greater than the costs, a minority of respondents felt that the opposite was true. Moreover, as indicated in the previous section, some positive impacts, such as the escalation in real estate prices, can be negative for some groups in the community. Whether one considers older in-migration to be hurtful or beneficial depends on one's perspective and place in the community's social structure.

An underlying dynamic of this situation is that older in-migration is changing these rural retirement communities' social stratification. We saw in Chapter 4, that older in-movers tend to have higher socio-economic status than longer-term residents in the destination communities. Their arrival tends to result in increased median income but also increased forms of inequality. Moreover, since the majority of older in-movers lived most of their lives in larger, more highly urbanized areas, they tend to expect a higher level of public service than is typically available in the RRDs to which they move. These differences in material resources, life experiences and expectations tend to set the stage for differences of opinion over the community's public agenda. Interviewees told us that older in-migration either was resulting, or might result, in housing displacement, cultural displacement, and political displacement.

Housing Displacement

Lack of affordable housing was identified as a migration-related problem in all four of the study communities. Housing displacement occurs when housing values increase at such a rate that certain groups cannot afford to purchase or rent housing in the community. In Lincoln, Maine, for example, we found that the in-movement of retirees into coastal towns, and subsequent rise in housing prices, forced many younger persons to move north of U.S. Route 1. Property values in the coastal towns south of Route 1, where many of these young persons were born and raised, had simply gotten out of reach. While some of these younger persons have benefited from new jobs in construction, hospitality or allied health services in the coastal towns, they can no longer afford to live near where they work. Many are forced to downsize their dwellings, double up with their parents or other relatives or move to a mobile home north of Route 1. A similar pattern of housing displacement was observed in Leelanau, Michigan, but rather than moving to the periphery of the county, young workers were displaced south into Benzie County. In Leelanau, the situation is further complicated by the fact that most new jobs in the county are seasonal. Many young persons lack the ability to earn a year-round income in their home county. As a result, these individuals move farther south to the state's metropolitan areas, or out of the state or region entirely.

This demographic shift has a major effect on public education in Leelanau because school aid in Michigan is allocated on a per pupil basis. Even though the county's overall population size has remained relatively stable, its age composition has grown older with the in-movement of older persons and the out movement of families with school age children. Accordingly, financial support for education in the county has declined. Older in-movers tend to support public education, but the county's current configuration of four independent school districts will probably be unsustainable in the future. Some form of consolidation or reorganization will be necessary. One interviewee also observed that the out-movement of younger families made recruiting volunteers for EMS and fire protection difficult. He felt that the county would have to hire full-time professionals for these positions in the future.

The lack of affordable housing and the displacement of younger families to the periphery were also mentioned by interviewees in Transylvania, North Carolina, and in Gila, Arizona. In these instances, the problem tended to be framed as lack of "work-force housing" and in particular housing for teachers, nurses, first responders, allied health providers and other service workers. In Transylvania, one interviewee who is involved in the planning process indicated that most new housing in the county was in the $200,000–400,000 range and that the county lacked both affordable housing and "elder friendly" housing. He indicated that the town had recently changed its zoning ordinances to permit co-housing options that would meet the needs of a wider range of socio-economic groups. An elected public official in Transylvania commented that "the greatest threat is that this will become a have/have not community" because the escalation of real estate prices is displacing entry-level teachers and other service providing professionals. In Gila, Arizona one respondent characterized the housing situation as "Aspenization, everyone who works here lives somewhere else." We were told that housing price escalation is driven by high property turnover. Older in-movers purchase one house and then sell it and move to another house in the county or elsewhere, with prices increasing with each transaction. We were also told that there is current discussion of developing an "affordable housing trust fund" in the county that would subsidize the cost of mid-range housing so that service providers, young professionals and emergency responders can afford to live near where they work. Housing has become so expensive in the county that it is even difficult to recruit young physicians and other health care providers to work at the Payson Regional Medical Center.

Cultural Displacement

A number of interviewees commented that in-movers do not respect traditional ways of doing things. We were repeatedly told that in-movers were impatient with local methods and that longer-term residents considered them to be arrogant. Interviewees in all four communities told us that in-movers "don't understand how we do things here." Even when in-movers and longer-term residents agree on issues, for example there is widespread recognition in all four communities of the need for affordable housing, there tends to be disagreement over how to approach the situation. One interviewee in Gila, Arizona captured the ambivalence felt by many longer-term residents toward in-movers. He told us that older in-movers come from all walks of life, bring rich experiences to the community, and are a definite community resource, "but it's a problem when they think they know everything."

Many longer-term residents also feel that in-movers are disrespectful of the community's cultural heritage. In Lincoln, Maine, for example, one interviewee complained that in-movers lacked respect for fishing and farming culture and that this was jeopardizing traditional resource-based industries. In a related comment, several interviewees in Lincoln observed that in-movers' strong emphasis on proprietary property differs diametrically from old timers open access approach. As a result, access to coastal resources has been severely restricted with obvious negative implications for fishing and other marine-based pursuits.

In both Lincoln and Leelanau longer-term residents were irritated by what they perceived as the in-movers' disrespect for traditional buildings and other structures. In both places interviewees told us that in-movers bought fine houses, only to tear them down and replace them with "McMansions." Locals felt that many of these new structures were unsuited to their communities. In fact, in-movers often replace two or more pre-existing houses with one large modern dwelling on a double lot. We were also told about disagreements between in-movers and longer-term residents over public art and artifacts. In Payson, Arizona, for example, one interviewee told us that in-movers wanted the town to "throw out all that old junk," and in Brevard, North Carolina, the sighting of public sculpture in the downtown business district was opposed by longer-term residents. In all four case study communities we heard that longer-term residents felt that gentrification was artificial and disrespectful of the community's architectural and aesthetic legacy. In contrast, we also saw evidence of cultural preservation of traditional lake-side villages such as "Fishtown" in Leelanau County, Michigan.

In-movers and longer-term residents differ in their attitudes toward further economic growth. While many longer-term residents are prone to label in-movers as *anti-growth*, we found that they are more likely to embrace growth management rather than "pulling up the gangplank." This is most clearly evidenced in Leelanau, Michigan where in-movers dominate the economic development commission (EDC) and are prime movers in land conservancy. Rather than opposing further growth and development, the EDC and the conservancy favor "smart growth." They promote zoning ordinances that would restrict new development to the villages and preserve agriculture, open space and "rurality" outside of the villages. The EDC sees retirement migration as the engine of growth in the county, but they want to carefully manage where it occurs. Several interviewees told us that longer-term residents see this as a violation of their property rights and "an attempt to put land away." Similarly, newcomers to Payson, Arizona, favor growth management. Their 17-point managed growth plan has brought them into political contention with the county's long-standing elite, ultimately resulting in the town's political realignment. Several interviewees confided to us that longer-term residents believe the 17-point plan is anti-growth. In contrast, the plan's proponents see it as an evidence-based approach to placing new development in the most appropriate and sustainable locations.

The land trust is also active in Lincoln, Maine, which has over a dozen preserves. Similar to Payson and Leelanau, many longer-term residents do not see the need to "put land aside." In addition, the pro- versus anti-growth sentiment plays out in other ways here. Rather than simply focusing on the amount and location of residential development, in-movers to Lincoln have led an organized campaign against big box development. As a result, five of the county's towns recently voted to ban big box development. Longer-term residents feel that this denies them access to jobs and affordable merchandise, while many older in-movers see big box development as unsightly.

Political Displacement

Many of the officials and leaders we interviewed told us that they were concerned that in-movers wanted to acquire political power and displace longer-term residents from political leadership. In Lincoln County, Maine one interviewee maintained that in-movers were attempting to "co-opt local government" and in particular take over the open town meetings where many local decisions are made. Ironically, longer-term residents in several towns are promoting a change to a referendum form of decision making because they feel that this more private process will help them maintain their political power. One interviewee, commenting on this situation, characterized it as a process of "re-colonization" where in-movers "wrested power from the natives." While this may be something of an exaggeration, it reflects a perception held by several of the longer-term residents we interviewed.

The situation in Lincoln is further complicated because several interviewees told us that longer-term residents and in-movers have different concepts about the role of government in local affairs. Longer-term residents were characterized as believing that informal social controls, by and large, are sufficient to govern the community. In contrast, many older in-movers were said to support a more activist role for government and to believe that formal rules and regulations are necessary to address issues such as the environment and land use. It should be noted that these in-migrant vs. longer-term resident differences vary across the towns that comprise Lincoln County and are much more salient in some locales, such as Westport Island, than in others.

In Leelanau, Michigan, in-movers influence public policy as members of various boards and commissions and through organizational participation. However, we did not perceive that locals feel threatened by this, nor do they see this as displacing longer-term residents from community power. In fact, participation by in-migrants on boards and commissions was considered to be beneficial for the local community by several of the persons we interviewed. However, as indicated in the earlier section on cultural displacement, in-movers to Leelanau tend to favor growth management and open land preservation. Conversely, longer-term residents see little need for "putting land away," and in-movers use their positions on commissions and boards and in non-profit organizations to promote their positions. This is seen by some longer-term residents as undermining the electoral process because in-movers, whose dwellings are not officially designated as their primary residence, are not permitted to vote or hold electoral office in Leelanau County. In contrast, some in-movers see this as "taxation without representation." As an in-mover who is active in the land trust told us, "organizational participation is a way to gain voice for people who own second homes in the county."

Several interviewees in Transylvania County, North Carolina, told us that older in-movers are a powerful political force in the county. One elected official said that seniors regularly attend council meetings, and "ninety nine percent of seniors vote." While older in-movers have not tended to seek elected office or official positions of power, their presence is still felt in the community. Similar to the situation in Lincoln, Maine, their agenda differs somewhat from that of longer-term residents. We were told that the community had traditionally had a strong libertarian flavor, but that in-movers want a more active and regulating government. One interviewee told us that "there are no native democrats in this county." This is undoubtedly an exaggeration, but it reveals how one key informant views the in-movers in comparison with persons who have lived in the area for a longer time.

We observed the clearest evidence of political displacement in Payson, Arizona. Similar to Transylvania, North Carolina, older voters are a powerful political force in the county. An elected official told us that "sixty-two percent of the voters in Payson are 55 or older," and many of these persons are in-movers. Many of the leaders and officials we interviewed told us that in-movers and longer-term residents were locked in stiff political competition in the county, with in-movers recently displacing longer-term residents from many elected and appointed offices. One in-mover who was recently elected to public office characterized the old guard as "lords." He told us that in-movers brought new, "more sophisticated," ways of thinking to the community. He also said that the in-movers formed a coalition with less well-off longer-term residents to unseat the old guard. As indicated in the earlier section on cultural differences, some longer-term residents interpret actions by in-movers as being anti-growth, while in-movers themselves see their actions as promoting sustainable development. Payson's new elected officials favor professional planning and growth management, often deploying the language of "smart growth" to justify their actions. In contrast, a longer-term resident we interviewed complained that in-movers do not respect traditional land use patterns. He reported that the in-movers "brought their own *environment* with them." When we asked one elected official whether there was conflict between in-movers and longer-term residents, he replied, "once you do away with the old power elite, there isn't much conflict." It remains to be seen how sustainable this situation is. Many persons we interviewed confided that there was a lot of political tension in the community.

Social Relationships between In-Movers and Longer-Term Residents

While in-movers and longer-term residents tend to hold different views on a number of critical issues, we did not observe any instances of outright conflict between the two groups. We asked interviewees explicitly whether in-migration generates conflicts in the community and the vast majority reported that there was contention but not conflict.

We also asked respondents whether "people around here consider the in-movement of older people to be an important issue." Most of the leaders and officials we interviewed said that it was an issue, something that locals talk about and something that is discussed in the local media. When we asked them to rank older in-migration in importance compared with other issues, however, most respondents indicated that older in-migration per se was not the fundamental issue. Rather, they said that it underlies other issues like affordable housing, land use, education and health care. Some persons commented that older in-migration results in a changed community context that has to be considered.

Most respondents felt that older in-movers "fit into the community." Even in counties where a substantial portion of older in-movers live in gated communities, we learned that they "come out from behind the walls and partic- ipate in the community." We asked leaders in service providing organizations whether in-movers and longer-term residents interacted within their organiza- tions. While there was some difference across the four communities, in general we found that service organizations tended to be spaces where at least a modest amount of social interaction occurred between older in-movers and longer-term residents.

We were intrigued by how longer-term residents refer to in-movers, and we believe that these labels reveal something about longer-term residents' attitudes toward their new neighbors. We found that longer-term residents refer to in-movers as a *generalized other*. This draws a bright boundary between the two groups and does not acknowledge any within-group variability. In Lincoln County, Maine, for example, in-movers were simply referred to as "people from away" or "PFA's." In Leelanau, Michigan, because local fudge is popular among tourists, in-movers were referred to as "fudgies" if they are summer tourists or "perma-fudgies" if they were seasonal residents. In-movers to Transylvania, North Carolina are referred to as "Yankees" regardless of whether they come from the North or from another region. More recently, many "half backs" have found their way to Transylvania after first moving from the Northeast to Florida, with their name stemming from the fact that they moved half-way back to where they originated before moving to Florida. Longer-term residents of Gila, Arizona, did not appear to have a particular label for in-movers. We speculated that this is because in-movers comprise such a large share of Payson's population overall, and because older in-migration is such a recent phenomenon here.

Throughout our interviews, in all four case study communities, we wondered whether in-migrant vs. longer-term resident differences resulted from social class differences or length of residence. Upon completion of the project, we concluded that both factors are important in structuring the social lives of older persons who move to rural retirement destinations (RRDs). As indicated in Chapter 4, in-movers, for the most part, are positively selected with respect to income, educational attainment and occupational prestige, and this contributes to the strengthening of class boundaries in the rural retirement communities. Moreover, as discussed in the previous section on politics, in-movers often ally themselves with better off segments of the destination community which tends to change the balance of power between better off and more economically modest population subgroups. In-movers themselves told us that most of their social contacts were with persons of similar socio-economic background, although they were not necessarily limited to interacting with other in-movers or with persons who lived close to them in the community. Much of their social contacts take place in the service organizations where they volunteer and in churches and other religious organizations.

Expectations built in previous places of residence also contribute to the production of social boundaries between in-movers and longer-term residents. Many of the leaders, officials and service providers we interviewed told us that in-movers had different expectations for government and community than was true of longer-term residents. We were told that longer-term residents and in-movers not only differed on what they felt needed doing, but on how to do it. One interviewee told us that in-movers want to live in a rural area with urban style services. Others told us that older in-movers were "extremely demanding." It was reported that in-movers expected the public sector to respond to their needs and demands because they pay a substantial amount in taxes. Many longer-term residents told us that they were ambivalent about the older in-movers. As one person confided, "we can't bite the hand that feeds us, but they are a pain in the ass."

CONCLUSIONS

The case study research that frames this chapter was motivated by three interdependent questions: (a) What aspects of community are most directly affected by older in-migration? (b) What contributions do older in-migrants make to local economy and society? (c) What is the nature of social relationships between older in-movers and longer-term residents? Our in-depth studies in four geographically and socially distinct RRDs showed that while generalizable answers to these questions are sometimes possible, important inter-community differences also exist. In other words, it is not possible to generate a one size fits all explanation of the challenges and opportunities associated with older in-migration to rural retirement destinations.

Our first observation is that communities do not mechanistically respond to changes in their population size and/or composition. Rather, local contingencies mediate these impacts. So, while a high rate of in-movement among older persons may generate increased demand and need for aging-related medical specialists and allied health personnel, an increased supply of such specialists does not automatically result. The impact of older in-migration on changes in the local health care sector is mediated by housing costs for service workers, the development strategy employed by local hospitals, institutional capacity to train indigenous workers and competition with nearby larger places.

We also observed that the advent of older in-migration to a rural retirement destination initiates a developmental process that plays out over time. We hypothesize that many institutional responses only occur after older in-migration has flowed to a community for a decade or longer. Moreover, the impacts of older in-migration seem to be contingent on whether the community remains a "summer place" or becomes a year-round community. Only in Lincoln, Maine, where older in-migration has been going on for decades, and which has been transformed to a year-round community, did we observe in-movement of older persons having a significant generalized effect on the local economy. We speculate that similar impacts are in store for the other areas once a sufficient number of in-movers have begun to age-in-place. This is consistent with our finding reported in Chapter 3 that older in-migration seems to be a self maintaining process and that an increase in percent 65 and older is one of the strongest predictors of whether a RRD will maintain its status over time or drop out of the category.

Aging-in-place was seldom discussed by the informants we interviewed, but it may ultimately be just as important a challenge as older in-migration. Housing and community infrastructure are being produced with the current population's needs in mind, and most older in-movers are healthy, mobile, independent and relatively well off economically. As they age-in-place, however, in-movers' housing needs may change dramatically. Moreover, most in-movers live in subdivisions located at a substantial distance from health care, commercial opportunities, leisure time activities and other necessities. This is fine as long as they continue to drive, but it will become problematic after the cessation of driving. We found very little evidence that any of the rural retirement communities we studied has a systematic strategy to meet this need. In fact, Payson recently rejected a study that proposed investigation of a mass transit system.

Our case studies indicate that the term *retirement in-migration* may miss an important aspect of the older migration phenomenon. We learned that many older in-movers either retain their jobs or rejoin the workforce once they arrive in the retirement destination. Some even establish new businesses in their new communities. Labor force re-entry seems to be motivated by a variety of both economic and social factors. Some older persons find that they underestimated the cost of living in their new homes and that supplemental income is required. Some persons have more than sufficient income but desire a job for the social contact and collegiality it provides.

Virtually every person we interviewed told us that older in-movers were important community resources. They donate money, but more importantly they have time to volunteer and professional and managerial expertise to apply to community problems. They are the main stays and prime movers of many organizations in RRDs. But like most of the other issues discussed in this chapter, volunteering is a complex behavior. For example, older in-movers provide their time, money and expertise to different types of organizations in some communities than in others. And volunteerism can be a double edged sword. While it provides cost saving labor and professional expertise, it may contribute to political contention between in-movers and longer-term residents and also create a secondary labor market undermining the demand for paid professional workers in the community. As a result, this may contribute to out-migration of better educated younger persons, increased inequality and a lower level of overall economic development.

While the bulk of this book focuses on older in-movers themselves, this chapter demonstrates the importance of understanding the social and economic context within which migration occurs. Virtually every issue we examined here is seriously under-researched. Volunteerism, in particular, merits additional systematic analysis. Finally, this chapter demonstrates the importance of combining various methods and types of data when conducting research such as this. Our case studies contain rich detail which complements our survey and census data analyses. Comprehensive examinations benefit from a diverse research strategy.

NOTES

1. Selectmen is a generic term for elected town-level officials in New England. This term refers to both men and women office holders.
2. Each interview commenced with a reminder that the interview was voluntary and with assurances that all information provided would be kept completely confidential. We also got written permission to tape the sessions and, in some instances, to take the respondent's picture for use in this book. Our procedures were approved by the Cornell University Human Subjects Research Committee.

3. In Arizona, places with 3,500 or more persons can choose to become a city (with a charter) or remain towns. Payson, with a population of 13,620 in 2000, has chosen to remain a town.
4. This is the smallest reservation in Arizona. The tribe is acquiring land from the U.S. Forest Service and expects to expand to 240 acres in the near future (Center for American Indian Economic Development, 2007).
5. There are also three unincorporated communities in the county.
6. While the percent of school-aged persons has decreased in Leelanau County, the number of school age persons has actually increased. This increase is strongly concentrated in several townships and in Suttons Bay, not where most of the educational infrastructure is located. Even though the county is resisting school consolidation, there has been some movement toward sharing of administrative staff and other inter-school district cooperation.
7. We also had informal conversations with business owners, restaurant workers and persons on the street in all four communities. These conversations are not systematically analyzed, but they were revealing in a number of ways. We will identify information gained in this manner where appropriate and useful in our narrative.
8. This is because of Michigan's Homestead Law.

REFERENCES

Brink, P. (2006). *Transylvania county, N.C. Newcomer's guide*. Brevard: The Transylvania Times.
Brown, D., Cromartie, J., and Kulcsar, L. (2004). Micropolitan areas and the measurement of American urbanization. *Population Research and Policy Review, 23*(4), 399–418.
Brown, D., and Glasgow, N. (1991). Capacity building and rural government adaptation to population change. In C. Flora and J. Christianson (Eds.), *Rural policies for the 1990s* (pp. 194–208). Boulder: Westview Press.
Center for American Indian Economic Development. (2007). *Tribe pages*. Flagstaff: University of Northern Arizona. Retrieved March 2007, from http://www.cba.nau.edu/caied/TribePages/Tonto.asp.
Fuguitt, G., Brown, D., and Beale, C. (1989). *Rural and small town America*. New York: Russell Sage Foundation.
Hummon, D. (1990). *Common Places*. Albany: SUNY Press.
Leelanau County. (2000). *Seasonal Population* (Working Paper Number 15). Leelanau County, Michigan.
Leelanau County. (2003). *Demographics* (Working Paper Number 11). Leelanau County, Michigan.
Lincoln County (2006). *Lincoln county information*. Retrieved November 2006, from http://www.co.lincoln.me.us/county.html
Massey, D., Alarcon, R., Durand, J., and Gonzalez, H. (1987). *Return to Aztlan*. Berkley: University of California Press.
Rim Country Chamber of Commerce. (2006). *Arizona Rim Country*. Payson, Arizona.
U.S. Census (2000). *Summary File 3 (SF3)*. Washington, D.C.: U.S. Census Bureau.
Warner, M. (2003). Competition, cooperation and local governance. In D. L. Brown and L. Swanson (Eds.), *Challenges for Rural America in the 21st Century* (pp. 252–261). University Park: Penn State University Press.
Warren, R. (1978). *The community in America*. Chicago: Rand McNally College Publishing Company.

APPENDIX

List of Positions Held by Interviewees in Four Rural Retirement Destination
Counties, 2006

	Government, Media, Business	Service Providers
Lincoln, Maine (7/12–7/17)[a]	County administrator	Co-chairs YMCA fund drive
	Director Office of Economic Development	Director, Senior Spectrum
	Editor, *Lincoln County News*	VP for development, Miles
	County commissioner	Health Care
	Editor, *Boothbay Register*	
	Selectman, Southport	
	Real estate developer	
	Selectman, Westport	
	Real estate broker	
Leelanau, Michigan (7/31–8/4)	Editor, *Leelanau Enterprise*	Director, Commission on
	Director, Planning and Community	Aging
	Development Office	Director, Friendship Senior
	President, Economic Development Corporation	Center
	Director, Chamber of Commerce	Administrator, Tender
	County administrator	Care assisted living
	Suttons Bay town planner	Director, Share Care
	Real estate broker	
Transylvania, NC (9/28–10/3)	County planner	Staff, Area Agency on Aging
	Editor, *Transylvania Times*	Owner, assisted living facility
	Brevard city planner	County agents NCSU
	President, Brevard Chamber of Commerce	Cooperative Extension
	Mayor of Brevard	
	Chair, County Economic Development Board	
	Director, SCORE	
Gila, Arizona (10/19-10-25)	Payson town manager	Receptionist, Payson Senior
	Editor, *Payson Roundup*	Center
	Director, Payson Community	Director, Gila County Aging
	Development Department	Services
	Director, Payson Economic Development	Director, Gila RSVP
	Department	
	Mayor of Payson	Administrators Payson,
	Director, Rim Country Chamber of Commerce	Regional Medical Ctr.

[a] Date visited in parentheses.

CHAPTER 7

RURAL RETIREMENT MIGRATION
AND PUBLIC POLICY

INTRODUCTION

In this book we have examined rural retirement migration from the older in-migrants' perspective and from the vantage point of the destination communities to which they move. This integrated micro–macro approach has permitted us to view older in-migrants as embedded in particular types of social environments that facilitate and constrain their opportunities for productive living during older age. It also permits us to examine the positive and negative effects of older in-migration for destination communities. In this chapter we will briefly review some of the main lessons learned from this research and then examine the implications that both these findings and the more general literature on retirement migration have for public policy. Like most social science research, our analyses answer some questions and raise others for further study. Accordingly, we close the book by identifying a number of high priority issues that merit future research.

WHAT WE LEARNED IN THIS RESEARCH

Our research focused on retirement-age migration to unplanned rural retirement destinations (RRDs). In the course of our research we found that the conventional distinction between planned and unplanned retirement communities is a false dichotomy to some extent. The communities we studied, while not explicitly developed for retirement, all included some measure of purposeful development for senior living.[1] For example, Brevard, North Carolina did not embark on an explicit strategy to replace manufacturing with retirement. However, Lake Toxaway and Connestee Falls, two areas of the county where many retirees live, are both gated communities with extensive recreation facilities that are provided exclusively for their residents. Similarly, Boothbay Harbor, Maine, did not set out to become a retirement destination, and while most retirees there do not live in comprehensively planned developments, some reside

179

in St. Andrews Village and other planned estates within Lincoln County. We also found that, while local governments do not typically market their communities as retirement destinations, local developers in these areas often have active recruitment campaigns and retiree attraction is an explicit government policy in some states.

Consistent with previous research, our study demonstrates that older in-migrants are positively selected with respect to socio-economic status. However, these positive selectivities are typically quite modest when older in-movers are compared with longer-settled residents of *similar age*. Our data showed that only four percent of the older residents of rural retirement destinations we surveyed were born in these places. We considered longer-term older residents of rural RRDs to be "non-migrants," but many of them are themselves in-migrants who moved in sometime in the not so distant past. Even though longer-term older residents have lived in our study communities for an average of over 20 years, they themselves tend to be positively selected compared with the general population residing in retirement destinations. Like other forms of geographic mobility, retirement migration follows well developed streams, and once established these streams become self sustaining. In other words, current older migrants tend to join previous in-movers who are aging-in-place. Previous research has tended to view older migration as an essentially individualistic behavior, but like all migration the geographic movement of retirees and other older persons is motivated and guided by social networks.

Our study provides new insights into the migration decision-making process among older persons and into the duration of residence older migrants can expect to have in rural retirement destinations. While our analysis confirms the previous characterization of migrants who move in their 60s as "amenity movers" (Litwak and Longino, 1987; Wiseman, 1980), we also show that migration decision making is a complex process that typically involves multiple motives. Respondents to our survey indicated that community atmosphere, landscape, weather and opportunities for outdoor recreation were important factors attracting them to their new communities, but over a quarter told us that family was the main reason they chose their new place of residence. In fact, our data show that more than one-third of in-movers had at least one adult child living within a half hour drive of their new residence. Hence, rather than simply being an "amenity-move" we believe that many older in-migrants to rural retirement destinations are combining environmental and community preferences with the consolidation of family ties. Older in-migrants want easy access to their children and grandchildren, and it seems that they are also being realistic about their need for assistance that will help them maintain an independent life as they age-in-place (Silverstein and Angelelli, 1998). In other words, moving closer to relatives enhances access to natural helper networks that contribute to independent living over the long-term.

This finding casts some doubt on the stage to stage progression described by Litwak and Longino (1987) in their developmental model of older migration. For some proportion of older in-migrants to rural retirement destinations we believe that individuals combine stage 1, "amenity-driven moves," with stage 2, "assistance moves," resulting in the potential for significantly longer durations of residence in rural retirement destinations than was previously suggested (see Figure 7.1). While many older in-movers can be expected to return to their origin communities after living in a retirement destination for a period of time, many others will spend the remainder of their lives in these rural retirement communities. This raises questions about the costs and benefits associated with retirement migration over the longer-term. As older in-migrants age-in-place their contributions to community are likely to decline in relationship to their increasing costs and needs. Family and friendship networks can reduce the need for publicly-provided assistance, but we wonder about the circumstances under which older in-migration might eventually result in a high fiscal burden. We return to this issue later in the chapter.

Figure 7.1. Retiree Who Moved to be Closer to Her Daughter and Son-in-Law, Now Aging-in-Place

We were pleased to find that older in-migrants appear to have little trouble establishing social relationships soon after they move to rural retirement destinations. In fact, our survey data show that in-migrants' frequency of participation in organizations, clubs and volunteer activities differs very little from that of longer-settled older persons and their informal ties with family and friends were also considerable. Accordingly, our concern that older persons might become socially isolated as a result of disrupting long time social relationships in the process of moving to a new community appear to be unfounded. While this is reassuring, our multivariate analysis provided only modest support for the hypothesis that socially integrated persons would experience better health than other persons with similar characteristics who lack strong social connections. In fact, informal social relationships with family and friends had virtually no effect on older persons' health in RRDs; and when they did have an impact it was weakly negative. Formal social participation, as hypothesized, had a clearer positive impact on older persons' health, but the results were still much weaker than anticipated. Interestingly, however, we found that religious participation only benefited longer-settled older persons while only in-migrants benefited from diverse organizational involvements.

While Chapter 5 focused on social participation at the individual-level, Chapter 6 examined social participation from the community's point of view. We found copious evidence that communities place a high value on the services and resources provided by older in-migrants. We were told that they were the driving forces behind many organizations and social movements, the prime sources of volunteers used by community organizations, a tax savings source of free professional, technical, and managerial assistance, and the donors of significant amounts of money to a wide variety of community causes and organizations. In contrast, our case studies revealed a degree of ambivalence toward older in-migrants on the part of longer-term residents because of their deep community involvement. While recognizing and appreciating their contributions, many longer-term residents reported concerns with what we have characterized as economic, cultural and political displacement. One of our overriding conclusions is that older in-migration benefits some people while disadvantaging others and one's evaluation of the costs and benefits associated with older in-migration depends on one's position in the community. What might seem unambiguously positive to a home developer or a real estate agent may be viewed as contributing to housing insecurity by a teacher, nurse or first responder. Of course, these goals are not necessarily in conflict. In Payson, Arizona, for example, affordable housing was part of the 17 point growth management plan proposed by the mayor in his election platform.

One of our goals in conducting this study was to explain why some rural places are more likely than others to become destinations for older in-migration.

While we were not fully able to answer this question, our research suggests that places develop into retirement destinations over time and that, once begun, the process becomes self-sustaining. Our multivariate analysis in Chapter 3 showed that places that are aging through older in-migration continue to attract older persons over successive decades and that previous status as a retirement destination was a strong predictor of being one at the present time. In other words, once retirement living has become established in a particular community, migration streams linking such places with a pool of potential older in-migrants develop and persist.

Unsurprisingly, many rural retirement destinations began as sites of tourism and recreation. As shown by our multivariate analysis in Chapter 3, dependence on recreation and tourism employment continues to be a strong predictor of being a rural retirement destination at the present time. Not only are older persons attracted by the leisure and quality of life opportunities such places provide, but retirees often have long-standing connections with particular locations where they and their families vacationed earlier in their lives. This process of long-term contact is sustained by a powerful dynamic that may extend across multiple generations. Our interviews with retirees in Boothbay Harbor, Maine, and Leelanau County, Michigan, for example, revealed that retirees' own children and grandchildren often spend significant amounts of time with them at their retirement homes during the summer. As a member of the Chamber of Commerce in Leelanau, Michigan commented, "These people are some of the most visited persons in the country." This suggests that older in-migrants' adult children may follow their parents to rural retirement destinations at the time of their own retirement. With large baby boom cohorts nearing retirement age, rural retirement destinations should consider the possibility that the next generation of older in-migrants may arrive in the near future.

Our research has begun to reveal how initial contacts between potential in-migrants and retirement communities are made, how these links persist over time, and how they motivate future in-migration. Retirement destinations may encourage older in-migration through promotional efforts, but it is more likely to be the result of social networks that develop between early in-migrants and their families, friends and neighbors who join them later on. As discussed above, consolidation of family relationships can be a basis for recruitment, as can friendships. For example, our community case studies indicated that vacation friendships are often quite durable and can be a basis for recruitment to rural retirement destinations in later life. We were also told that vacationers often receive the summer place's newspaper during the winter months which maintains contact, stimulates interest in local issues and provides information about real estate prices and housing availability.

Rural retirement destinations appear to have a "life history." As indicated above, many begin as vacation destinations which "hollow out" during the winter. The first waves of retirees tend to be seasonal residents, especially in cooler climates. However, our research indicates that if retirement settlement persists for a long enough time, seasonal places may be transformed into year-round communities. Community leaders and business owners in Boothbay Harbor, Maine, a long-established "summer place," reported that many more businesses and services now remain open throughout the year. Another result of the reduction of seasonal population fluctuation is that the Boothbay Harbor area now has many more doctors, medical facilities, allied health providers and year round commercial establishments than was true in the past. If Boothbay Harbor is the most established retirement destination we studied, Payson, Arizona is the least, as it has only begun to attract large numbers of retirees for over the past decade.[2] Business leaders and public officials in this community told us that while older in-migration was having some noticeable effects, it had yet to result in much locally-owned commercial or retail activity in the town. Similarly, Leelanau Michigan is still primarily a "summer place" and many older in-migrants spend the winter months in warmer climates. As a result, few businesses stay open year round and the county has a shortage of year round medical care providers. If the community development trajectory we observed in Boothbay Harbor is predictive of a similar process in other areas which have begun to attract retirees, one wonders if Payson and Leelanau, for example, will develop more fully and become less seasonal as the next generation of retirees joins their parents and friends in these places in the future.

RETIREMENT MIGRATION AND PUBLIC POLICY

As indicated above, our research shows that older in-migration to rural retirement destinations often combines "amenity-driven moves" and "assistance-driven moves" into a single residential change that occurs in the early to mid-60s. As a result, a substantial share of older in-migrants are likely to spend their remaining years in rural retirement destinations rather than returning to their origin communities. This has important implications for the in-migrants themselves and for the communities in which they have chosen to live. At the individual level Riley and Riley (1994), Glasgow (2004), Carp (1988) and others have observed that successful aging is strongly affected by the "person-environment fit." Similarly, from the community's perspective, viability is affected by the number, characteristics and well-being of older persons, as well as their civic engagement and social participation (Le Mesurier, 2006; Stallman, Deller and Shields, 1999; Rowles and Watkins, 1993). Even though the individual- and community-level implications of older in-migration may be separated for analytical purposes, in reality they are mutually intertwined. In other words, promoting successful aging benefits both older people and their communities.

Individual-Level Policy Considerations: Promoting Successful Aging

If we are correct that a substantial share of older in-migrants to rural retirement destinations will age-in-place, then these persons will benefit when communities take a longer-term view of the structural conditions that either impede or facilitate older persons social involvement. Previous research has shown that social integration is a powerful determinant of health and well-being among older persons (Thoits and Hewitt, 2001); and as demonstrated by Young and Glasgow (1998), organizational involvement appears to be particularly important. While our examination of this hypothesis in Chapter 5 provided only modest support, our findings were consistent with past research. Moreover, our face-to-face interviews showed that older in-migrants experience a multitude of benefits from being socially connected even if the connection with their health is not as strong as expected (Chapter 6). This is consistent with Wethington, Moen, Glasgow and Pillemer's (2000) observation that older persons' well-being is contingent on maintaining multiple roles, social support and social integration. As we discussed in Chapter 5, primary social relationships provide older persons with social support and buffer them from difficult situations such as the onset of disability, cessation of driving or financial reversals. Participation in formal organizations embeds them in problem-solving environments which provide important information as well as bridging ties to a wider set of social engagements (Young, 2004).

How can communities intervene to enhance social participation among older persons as they advance in age? As Riley and Riley (1994) have commented, this involves changing social attitudes toward older people as well as promoting opportunity structures that recognize rather than deny older people's productive capacities. What concrete actions can be taken? First, community leaders can promote a normative environment that values its older citizens and rejects the myths of ageism that tend to stereotype older persons as a class of unproductive individuals who occupy an "empty role" (Pillemer and Glasgow, 2000; Rosow, 1967).

From a structural community perspective, providing safe, efficient and affordable transportation is one way to facilitate continued social participation among older persons. Geographic mobility is critical for retaining social ties to the larger community and for occupying multiple social roles. Previous research has shown that most older persons drive themselves to the store, the doctor and to social engagements, and hence the public sector has a relatively small role in facilitating their continued social involvement and civic engagement. However, as Glasgow (2000) has demonstrated, driving transitions can put older persons at risk of social isolation. While 85 percent of current older drivers in her New York survey reported that they were able to go places as often as they liked, only about one half of former drivers reported being able to do so. The cessation of driving is a critical life course transition that will take on added salience as successive cohorts of older persons reach advanced ages (Burkhardt and Berger, 1997).[3]

Communities can enhance geographic mobility among older non-drivers and drivers alike by providing both formal and informal alternatives to driving. While many communities have experimented with creative trans-portation solutions, rural communities are much less likely to provide trans-portation than their urban counterparts (Cordes, 1985; Coward, Bull, Kukulka and Galliher, 1994). Accordingly, if rural older non-drivers are unable to obtain rides from friends and family, they are at high risk of being "transportation disadvantaged" and thereby socially isolated. Our research shows that about one third of older in-migrants to rural retirement destinations have at least one adult child living within a one-half hour drive; and, in principal, these persons should be willing and able to drive their aging parents and friends should they relin-quish their licenses or begin to limit their driving at night and in foul weather. If older persons' natural helper networks are disrupted, however, social isolation becomes a real possibility.[4] This is where the public sector will need to step in and provide formal services to substitute for the loss of own-driving and/or rides from friends and family. Unfortunately, our interviews with public officials in four rural retirement destinations did not indicate that planning for older residents' future needs in general, or for their transportation needs in particular, was a prominent issue on the public agenda. Perhaps this will change as more of their older residents age-in place, but current indications do not suggest that this is highly probable.

Community-Level Policy: Equitable Rural Development

Attracting retirees is widely considered to be an effective rural devel-opment option (Reeder, 1998). Many states have adopted explicit policies to attract older in-migrants and, while localities are less likely to actively recruit retirees, developers in such places often produce and market residential products targeted to well-off older persons. Proponents of such strategies point to demographic research that shows that retirement destinations have grown more rapidly and consistently over the last 30 years than other types of rural areas (see Chapter 3; Cook and Mizer, 1994), and economic research that shows that the benefits of retirement attraction exceed its costs (Glasgow and Reeder, 1990; Serow, 2003; Stallman, Deller and Shields, 1999). The policy question, however, is whether the research base on retiree attraction policies is sufficient to merit the investment of public resources in such schemes.

Chapters 2 and 6 contribute to this discussion by examining some of the social and institutional implications of retirement in-migration. Our focus on social factors in these chapters does not discount the importance of economic and fiscal issues. Our goal, rather, is to widen the development debate by elevating social factors and promoting their consideration in local development discourse. We discuss three social policy themes in this section that are relevant for considering retiree-attraction as a rural development strategy. Moreover, since growth and development are not synonymous, we encourage the reader to consider the circumstances in which retirement in-migration might result in increased equity within a community rather than contributing to increased social inequality and inter-group cleavage. In order to do this, we contend that policy makers should: (a) consider the pros and cons associated with older in-migration from multiple standpoints; (b) realize that adaptation to changes in population size and composition are mediated by local social structure; and (c) employ a long-term view in planning for the impacts of older in-migration.

The Importance of Considering Multiple Standpoints

As indicated earlier, one of the overriding conclusions we draw from our research is that *social changes associated with older in-migration benefit some groups and disadvantage others.* Hence, whether one considers retiree in-migration to be beneficial or deleterious depends on where one is situated in the community. For example, our survey analysis and community case studies confirmed that older in-migrants provide important volunteer services to their new communities, but we also found legitimate concerns associated with the utilization of older volunteers. Virtually every local leader and official we interviewed testified that older in-migrants are a "community asset." In-migrants bring money, talent, time, experience, connections and professional and technical expertise to the community. Not only are in-migrants active in a wide range of social, cultural and service domains, they are often "the movers and shakers" of these movements and organizations. We were told that in-migrants play major roles in service organizations such as the YMCA, RSVP, the hospital auxiliary, meals on wheels, and the senior center. Additionally, they frequently contribute to arts and cultural groups and to churches and other religious organizations. One official in Brevard, North Carolina characterized older in-migrants as "grey gold." On this basis, it would seem that the social benefits of older in-migration can be added to economic benefits, further recommending retirement attraction as a rural development option.

In contrast, we learned that older in-migration can pose some important challenges that must be weighed against the aforementioned benefits. For example, while older in-migration stimulates the housing market and induces employment in construction and housing-related businesses, it also inflates housing prices, often displacing middle income tradesmen, first responders, teachers and other service providers from rural communities. Since this displacement is selective by income and occupation, it has a strong effect on social stratification and contributes to growing social inequality locally.

Many of the elected and appointed officials we interviewed told us that retirees provided accounting, legal, engineering, architectural and other services free of charge, thus saving tax payers large amounts of money. While this seems like an unambiguous benefit to the community, some of the officials we spoke with wondered whether it might undercut the demand for paid professional labor thereby making their communities less attractive migration destinations for younger professionals. Some were particularly concerned that this might reduce their communities' ability to retain talented young professionals who grew up locally and went away to college for their education.

We also found evidence that older in-migration cuts in both positive and negative directions when it comes to local culture. As indicated earlier, in-migrants are often the prime movers of local arts and cultural organizations, but some persons we interviewed confided that in-migrants were disrespectful of the community's cultural heritage. They felt that in-migrants were usurping the cultural agenda, replacing local tastes and preferences with those imported from elsewhere. In addition, we received widespread reports that in-migrants did not respect traditional ways of doing things, were arrogant and "think they know everything."

Similarly, our case studies revealed that in-migrants were displacing longer-time residents in the political arena in some localities. In Payson, Arizona, for example, in-migrants unseated the long standing power structure, winning many of the town's elected offices. But the process of public contention extends beyond electoral politics to the broader arena of community decision making. As shown by studies of community power structure (Hunter, 1963), control of the public agenda is of utmost importance in local policy and decision making. Many of the persons we interviewed told us that longer-term residents and in-movers were locked in stiff competition over what issues merited public consideration and action. One public official we interviewed characterized the situation as "re-colonization" where in-movers "wrested power from the natives." In Leelanau County, Michigan, in-migrants championed the land trust while longer-term residents characterized this as "putting land away" and did not consider it a legitimate issue for public intervention. Similarly, in Payson, Arizona, in-migrants promoted a 17 point growth management plan in order to "place development in the most appropriate and sustainable locations." Many longer-term residents, in contrast, interpreted this as anti-growth not growth management. It should be noted, however, that it is not clear whether this observed public contention is generalizable across communities and/or time. For example, research on rural in-migration during the 1970s found little evidence of anti-growth sentiments among adults who were urban to rural migrants (Fliegel, Sofranko and Glasgow, 1981; Sofranko, Fliegel and Glasgow, 1983). We are not sure whether the differences between our study and those reported earlier are explained by the different time, situation or a combination of the two.

The research reviewed above indicates that leaders in retirement destination communities need to consider the pros and cons of public actions when charting their communities' future directions. To insure that pros and cons are fairly balanced, the planning process should be transparent and inclusive because persons differ in their evaluation of the merits of public actions.

Adaptation to Population Change is Mediated by Social Structure

While socio-demographic transformations such as population aging can induce changes in community institutions, these impacts of are *not automatic*. They are mediated by local social structure and by the larger macro-structural and policy environments in which localities are embedded (Brown, 2008f). To assume that a unit change in population size or age composition automatically and mechanistically results in a similar magnitude of change in factors such as economic activity, medical care provision, and public expenditure on transportation, is to deny the agency of actors and the instrumentality of community institutions. For example, research has shown that public revenue and expenditure changes in response to population change are contingent on differences in community economic base (Walzer and Deller, 1991). Community sociologists have also commented on the process of adaptation to population change, observing that the accomplishment of community goals depends on the nature of social relationships within local populations. In other words, similar needs in different places do not necessarily translate into similar outcomes. Flora and Flora (2003), for example, indicate that communities vary in their stock of social capital, which can affect their responses to acknowledged social needs including the provision of new services for an aging population. Luloff and Bridger (2003) argue that accomplishing narrowly specified social goals, such as transportation for older persons, depends on a wider community engagement across specific issue areas.

Our research in Chapter 6 showed that similar rates of older in-migration have had different effects in destination communities depending on their respective histories, the length of time they have been attracting retirees, whether retiree settlement was seasonal or year round, the political climate and a host of other community attributes. For example, relatively rapid in-migration at age 60 and above to Boothbay Harbor, Maine was accompanied by expanded health care infrastructure while the lone hospital in Leelanau County, Michigan, which "hollows out" in the winter, recently closed. Another example is that while in Payson, Arizona, only fast food franchises and big box chain stores have set up shop, Brevard, North Carolina, a long time center for music and the arts, has a vibrant downtown full of locally-owned shops, boutiques, restaurants and services. Population change poses both challenges and opportunities for local communities and neither will be realized without concerted public action. The bottom line is that while in-migration at age 60 and older may imply that the need for particular kinds of services is growing, that housing and community design must adapt to older persons' needs, and that the demand for public transportation will increase, more older persons and/or a higher percentage at the older ages can have a variety of results depending on local community organization and history. Regardless of the pace of older in-migration, communities will not invest in aging-related services and programs unless the merits of such public interventions are widely appreciated throughout the community, unless the political establishment is willing to spend its "political capital" on aging-related projects and unless proposed projects are judged to be economically feasible.

Changes Associated with Population Aging Occur Over a Long Time Horizon

As indicated above, while many older in-migrants will return to their origin communities, a substantial proportion are likely to spend their remaining years in rural retirement destinations. In addition, if history is a guide, about nine percent of aging baby boomers will change residence and a disproportionate share of these movers will be guided by their preferences for rural living and by long established social ties to join earlier cohorts of older in-migrants in rural retirement destinations (see Chapters 1 and 3). Accordingly, communities will be challenged with planning for the needs of previous in-migrants who are aging-in-place at the same time that they must consider the needs of younger retirees who have arrived more recently. This means that the growing older population in rural retirement communities will become more diverse with respect to age, health, functional limitations and social networks. Accommodating the needs of this increasingly diverse *older* population will be challenging.

Being a rural retirement destination has generally been framed in a positive light because older in-migrants typically arrive as couples, in good health, and they are relatively well off (see Chapter 2). In addition, as shown in Chapter 5, in-migrants quickly establish social connections, appear to have little difficulty becoming socially integrated and frequently contribute to the community as volunteers. While local leaders realize that costs are associated with an aging population, they tend to believe that benefits will more than offset these expenses, and economic analysis tends to support their view, at least in the short-run (Glasgow and Reeder, 1990; Stallman, Deller and Shields, 1999).

Consistent with Litwak and Longino's (1987) developmental perspective on older migration, however, this view assumes that older in-migrants will leave rural retirement destinations when they experience adverse situations such as the loss of their spouse, serious illness or the onset of disability. In other words, community leaders may be "betting" that older in-migrants will only remain as long as their contributions to the community exceed their costs. But our research suggests that a substantial proportion of older in-migrants, especially those who have joined their adult children, will remain for the duration of their lives. Some of these persons will have sufficient savings and investments to purchase services from the private sector, others will not. In other words, it is likely that some older in-migrants will remain after their costs exceed their benefits. While this shifting cost/benefit calculus tends to be carefully considered in planned retirement communities, and especially in continuing care retirement communities, it does not seem to be on the public agenda in the unplanned retirement destinations that we studied in this research. Considering transportation, for example, we found that few communities were planning for the transportation needs of their aging populations even though they realized that as in-migrants age in the community, many will relinquish their driver's licenses and become dependent on family, friends *and the local government* for geographic mobility. In fact, Payson, Arizona, recently completed a transportation study but decided not to act on its recommendations. This situation is further exacerbated by the dispersed pattern of settlement in rural retirement communities. While some older persons live in neighborhoods with easy access to commercial, medical and social opportunities, many others have settled far from the community center in naturally attractive, but remote and hard to reach locations. Of course, transportation is only one example of a public service where need will expand and where costs will increase as older persons age-in-place.

While communities should take advantage of short- and medium-term benefits associated with retirement migration, these economic and social pluses need to be realistically balanced against future trends in public sector expenditures that will undoubtedly increase over time. This raises the question of *sustainability*, or how retiree-attraction can benefit the community in both the present and future generations (Skelley, 2004). One way to examine this question is to determine how the stream of costs and benefits associated with older in-migrants may change over time as in-migrants age-in-place, and what costs and benefits can be expected from newer cohorts of in-migrants who are arriving now. In other words, when communities plan for the future they need to consider the varying cost and benefit streams associated with younger-old and older-old in-migrants as an integrated package that will change over time.

Local governments in the U.S. appear to be hesitant to plan for population aging, but some other countries have gotten the message. For example, the Australian Local Government Association (2005) has developed a series of publications that discuss the importance of timely action by local government in response to population aging. They promote integrated local area planning (ILAP), or what is known as the "whole of council approach." ILAP encourages better integration and coordination of planning efforts both within and between levels of government to provide better services and facilities for aging communities.

Recently, a consortium of five national organizations has recognized the need for U.S. cities and counties to better meet the needs of their aging populations.[5] Under the leadership of the National Association of Area Agencies on Aging, with a grant from the MetLife Foundation, this project conducted a survey of 10,000 local governments to determine their "aging readiness." Their initial report, *The Maturing of America—Getting Communities on Track for an Aging Population*, examined the availability of aging-related programs, policies and services, while also focusing on local governments' ability to harness older adults' talents, skills and wisdom (National Association of Area Agencies on Aging, 2006). A second phase of the project now underway focuses on identifying "promising practices" to help cities and counties increase their capacity to effectively serve older residents.

SOME HIGH PRIORITY RESEARCH NEEDS

Like all social science research, we have answered some of the original questions that motivated our project, left others unanswered and raised new questions for future research. In the course of our study, we identified a number of high priority questions that merit additional investigation. While some of these questions focus on the process of aging at the individual level, others concern the communities where older in-migrants reside. But as indicated earlier, people and the places are mutually interrelated, and these micro–macro relationships themselves are poorly understood.

What Factors Predict Duration of Residence Among Older In-Migrants in Rural Retirement Destinations? How Does Community Participation Change Over Time as In-Migrants Age-in-Place?

Litwak and Longino's (1987) developmental theory of older migration has been extremely influential in portraying older migration as an aspect of the latter part of the life course. Consistent with their theory, it is undoubtedly true that many older in-migrants to rural retirement destinations will return to their origin communities, but our study indicates that a substantial proportion of older in-migrants will age-in place in the countryside. Our study is one of the first to follow older in-migrants after their arrival in rural destinations, but our three year window of observation barely scratched the surface. In our opinion, longitudinal research of the aging-in-place of older in-migrants is the highest priority area for future research within the overall domain of retirement migration. What factors determine why some older in-migrants return to their origin while others stay in destination communities? How do older in-migrants' social and institutional participation change over time as they age-in-place? How can the costs and benefits associated with older in-migration be measured over the latter part of the life course?

How are Rural Retirement Destinations Established and Maintained?

Our county-level multivariate examination of why some places are more likely to be or become rural retirement destinations was a first step toward unraveling this question, but much more research is required to fully under-stand the issue. In particular, we wonder why natural amenities failed to predict the probability that some rural counties are more likely than others to be or become retirement destinations. The natural amenity scale used in Chapter 3 is a strong predictor of overall rural population change (Johnson and Cromartie, 2006; McGranahan, 1999), and older in-migrants cite environmental amenities among their major reasons for choosing to move to a rural retirement destination community. Yet why did the amenity scale not predict being or becoming a RRD in our county-level analysis? This raises a more general methodological and conceptual issue regarding the merging of micro-and macro-level explanations in social research. In our case, the question is how can individual-level reports of reasons for destination choice be merged with aggregate-level data on the variability of place characteristics to better understand the development process that produces and maintains communities as retirement destinations? More research is required to explain why places become established as retirement destinations and why they continue to attract older in-migrants over time or cease being destinations for older in-migration. This question will take on added significance as the baby boom enters older age and more retirees are available to change their residences

What is the Causal Process Linking Population Aging and being a Retirement Destination?

The multivariate analysis in Chapter 3 raised a puzzling question about the causal process that underlines the positive relationship between county-level population aging and maintaining status as a rural retirement destination. We interpreted the strong relationships between percent 65 and older, and increases therein, with retaining RRD status as indicating that once a county is established as a place of concentrated retirement living it becomes highly likely to maintain this status. We argued that inter-generational social networks are established between older residents of retirement destinations and family and friends who they encourage to join them there later in life. Since most of our analysis of this question utilized county-level census data, we only have direct evidence from four case studies that social networks develop, that parents recruit their friends and children, and that persistence as a retirement destination depends on these types of relationships. We recommend that future examination of this causal puzzle utilize a multi-method approach where in-depth case studies, household surveys and census analysis are carefully integrated from the research's inception.

Does the Relationship Between Social Integration and Well-Being During Later Life Depend on Where One Lives and/or the Kinds of Organizations in which One Participates?

While a host of previous research demonstrates that social integration is strongly associated with health and well-being among older persons (Pillemer, Moen, Wethington and Glasgow, 2000), and that organizational participation is particularly effective (Young and Glasgow, 1998), our study finds only modest support for this relationship in the context of rural retirement destinations. We are reasonably certain that our modest findings result from the relatively brief interval of time between the two waves of our panel survey, and that a longer interval would have produced both more changes in health status and greater variability among respondents. It can also be argued that our measures of health could have been more robust and that more sensitive measurement would produce stronger results.

Methodological issues aside, however, our modest results indicate that the social integration-health hypothesis should be reconsidered and subjected to additional rigorous evaluation. In particular, we recommend that future research look more closely at the finding that close personal relationships either have no effect on older persons' health or diminish it. We have observed that close personal relationships are full of ambivalence, obligation and stress which accounts for the lack of a positive impact. This causal pathway requires more in-depth analysis using both quantitative and qualitative approaches. Similarly, the proposition that older persons gain information and additional social connections as a result of participating in organizations, and that both of these benefits result in improved well-being, also requires additional analysis. The fact that only in-migrants were shown to benefit from diverse organizational involvements, and that religious participation only affected non-migrant health indicates that more attention should be given to the kinds of organizational participation which offer older persons the greatest benefits. Perhaps it is not participation per se, but rather participation in especially salient organizations that matters most.

Moreover, because older persons are embedded in diverse social environments which vary in the opportunities they provide for social participation, we recommend that future research on social integration and health include a suite of contextual variables permitting one to examine the hypothesis that social participation among older persons, and the effects of such involvement, may be contingent on variability in social context such as rural vs. urban areas, RRDs vs. natural decrease counties, isolated rural areas in comparison with the urban fringe, or rural communities in states with dramatically different social welfare systems.

How Does Community Structure Affect Opportunities for Voluntary Social Participation among Older In-Migrants?

As reported in Chapters 2 and 6, older in-migrants have little difficulty becoming involved in community activities; and, once involved they often become the mainstays of organizations and movements. One important unanswered question related to volunteerism is why older in-migrants are more likely to volunteer for different types of organizations in different places? In other words, how does community structure facilitate or constrain older in-migrants' opportunities for volunteering? While focused explicitly on volunteering, this question speaks to the more general issue of structure and opportunity raised by Riley and Riley (1994). Our research identified examples of how variability in the local context affects the nature of voluntary participation; but a more general, theoretically-informed analysis is needed. For example, we found that in-migrants were more likely to volunteer for organizations catering to limited income persons in Payson, Arizona, than in other places because relatively few public programs for low income older persons exist in the State. In contrast, in Boothbay Harbor, Maine, where the state's social welfare system is more developed, recruiting volunteers for the senior center and other organizations that assist limited income persons was quite difficult. Boothbay Harbor's retirees were more likely to be involved in culture, the arts, or youth-oriented organizations like the YMCA. Our case study in Leelanau County, Michigan provided another example of how community structure affects the nature of in-migrants' volunteer activities. In this case we were told that many older in-migrants participated in social movements like the land conservancy, partly because Michiganders are only permitted to vote at their primary place of residence. Since many older in-migrants maintained primary residences elsewhere, participation in social movements was seen as a way of exercising a political voice outside of electoral politics.

What Demographic and Socio-Economic Dynamics Account for the Simultaneous Occurrence of Natural Decrease and Retirement In-Migration in the Same Areas?

Our analysis in Chapters 2 and 3 revealed the surprising finding that half of all rural retirement destinations also have natural decrease (ND), or more deaths than births. This is surprising because, as a category, natural decrease counties are considerably older than RRDs and natural decrease is caused by an older age structure, not low fertility rates among persons of childbearing age. Researchers and policy makers have typically drawn a bright distinction between natural decrease counties, which are seen as declining places, and RRDs, which are considered to be among the most vital rural environments in the nation (see Table 2.5). The fact that half of RRDs have more deaths than births casts this distinction and its interpretation into serious question. Virtually nothing is known about the social or economic situation in areas which simultaneously attract older in-movers and at the same time have more deaths than births. Is the social and economic environment in these "overlap" counties fundamentally different from the majority of natural decrease counties? How do RRDs that have natural decrease compare with other RRDs where fertility is sufficient to produce natural increase? In other words, are RRD/ND overlap counties on a different (poorer) development trajectory than RRD counties that also have natural increase?

Do RRDs Follow a Linear Development Path?

We have observed that RRDs seem to follow a developmental trajectory that is initiated when they begin attracting relatively large numbers of retirees as seasonal migrants (as we saw in Payson, Arizona, and Leelanau, Michigan) and culminates when retirement living becomes year round for most older in-migrants (as appears to be the case in Boothbay Harbor, Maine). A coherent theoretically-informed framework specifying RRD life history as a series of developmental stages would guide research on RRD community development. One important issue, however, is whether the trajectory experienced by one place can be expected to describe development over time in other places, or if place-specific experience is purely idiosyncratic, the result of factors such as individual place history, climate, natural amenities, state government policy. This framework could be used to examine why some places with similar characteristics and histories begin to attract older in-migrants while others do not, and what factors determine the transition from stage to stage after places become established concentrations of retirement living. In addition, it would be important to understand how the transformation of retirement living from seasonal to year-round affects other aspects of community organization, such as the availability of health care, the filling out of commercial structure, community politics and the community's physical design.

While this study has contributed much to the knowledge about retirement migration, many important questions remain unresolved. As we have indicated throughout this book, retirement migration is a world-wide phenomenon that will grow in importance during the next decades as the baby boom generation advances through the life course. The research reported in this book provides a factual basis to inform public policy as communities consider how to promote both successful aging and successful and equitable community development. We hope that our research helps to inform aging and community development policy, and we encourage other social scientists to consider examining some of the questions we have left unanswered.

NOTES

1. These are in contrast to comprehensively planned retirement communities like *Leisure World*, California, *Sun City*, Arizona, or one of the numerous Del Web developments.
2. In this case we are referring to year round residence, not wintering at the RV parks near Roosevelt Reservoir.
3. About nine out of ten persons between ages 65 and 74 reported driving in Glasgow's (2000) survey, but by age 85 this figure declined to 79 percent among men and 42 percent among women.
4. Natural helper networks can be disrupted by death of a spouse, relative or friend, out-migration of a relative or friend and/or diminished intensity of friendship.
5. The five organizations are the National Association of Area Agencies on Aging, the International City/County Management Association, the National Association of Counties, the National League of Cities and Partners for Livable Communities.

REFERENCES

Australian Local Government Association (2005). *Awareness to action: Local government's response to population ageing*. Canberra: Australian Local Government Association.
Brown, D. (2008f). Rural America as seen through a social-demographic lens. In J. Wu, P. Barkley, and B. Weber (Eds.), *Frontiers in resource and rural economics* (pp. 3–23). Washington, D.C.: Resources for the Future Press.
Burkhardt, J., and Berger, A. (1997 January). *The mobility consequences of the cessation of driving*. Paper presented at the Transportation Research Board annual meeting, Washington, D.C.

Carp, F. (1988). Significance of mobility for the well-being of the elderly. In the Transportation
    Research Board's, *Transportation in an aging society: Improving mobility and safety for
    older persons* (pp. 1–20). Washington, D.C.: National Research Council.
Cook, P., and Mizer, K. (1994). *The revised ERS county typology: An overview* (Rural Development
    Research Report No. 89). Washington, D.C.: U.S. Department of Agriculture, Economic
    Research Service.
Cordes, S. (1985). Biopsychological imperatives from the rural perspective. *Social Science and
    Medicine, 21*(10), 1373–1379.
Coward, R., Bull, C., Kukulka, G., and Galliher, G. (1994). *Health services for rural elders.*
    New York: Springer.
Glasgow, N. (2004). Healthy aging in rural America. In N. Glasgow, L. Morton and N. Johnson
    (Eds.), *Critical issues in rural health* (pp. 271–281). Ames, Iowa: Blackwell Publishers.
Glasgow, N. (2000). Transportation transitions and social integration of nonmetropolitan older
    persons. In K. Pillemer, P. Moen, E. Wethington and N. Glasgow (Eds.), *Social integration
    in the second half of life* (pp. 108–131) Baltimore: Johns Hopkins University Press.
Glasgow, N., and Reeder, R. (1990). Economic and fiscal implications of non-metropolitan retirement
    migration. *Journal of Applied Gerontology, 9*(4), 433–451.
Fliegel, F., Sofranko, A., and Glasgow, N. (1981). Population growth in rural areas and sentiments
    of the new migrants toward further growth. *Rural Sociology, 46*(3), 411–429.
Flora, C., and J. Flora. (2003). Social capital. In D.L. Brown and L. Swanson (Eds.), *Challenges
    for rural America in the twenty-first century* (pp. 214–227). University Park: Penn State
    University Press.
Hunter, F. (1963) *Community power structure: A study of decision makers.* Garden City, NY:
    Anchor Books.
Johnson, K., and Cormartie, J. (2006). The rural rebound and its aftermath: Changing demographic
    dynamics and regional contrasts. In W. Kandel and D.L.Brown (Eds.), *Population change
    and rural society* (pp. 25–50). Dordrecht: Springer.
Le Mesurier, N. (2006). The contributions of older persons to rural community and citizenship. In
    P. Lowe and L. Speakman (Eds.), *The ageing countryside: The growing older population
    of rural England* (pp. 133–146). London: Age Concern England.
Litwak, E., and Longino, C. (1987). Migration patterns among the elderly: A developmental
    perspective. *The Gerontologist, 27*(3), 266–272.
Luloff, A., and J. Bridger. (2003). Community agency and local development. In D.L. Brown and
    L. Swanson (Eds.). *Challenges for rural America in the twenty-first century* (pp. 203–213).
    University Park: Penn State University Press.
McGranahan, D. (1999). *Natural amenities drive population change* (Agriculture Economic Report
    No. 718). Washington, D.C.: U.S. Department of Agriculture, Economic Research Service.
National Association of Area Agencies on Aging. (2006). *The maturing of America: getting commu-
    nities on track for an aging population.* Washington, D.C.: National Association of Area
    Agencies on Aging.
Pillemer, K., Moen, P., Wethington, E., and Glasgow, N. (Eds.). (2000). *Social integration in the
    second half of life.* Baltimore: Johns Hopkins University Press.
Pillemer, K., and Glasgow, N. (2000). Social integration and aging: Background and trends. In
    K. Pillemer, P. Moen, E. Wethington and N. Glasgow (Eds.), *Social integration in the
    second half of life* (pp. 19–45). Baltimore: Johns Hopkins University Press.
Reeder, R. (1998). *Retiree attraction policies for rural America* (Agriculture Information Bulletin No.
    741). Washington, D.C.: U.S. Department of Agriculture, Economic Research Service.
Riley, M.W., and Riley, J. (1994). Structural lag: Past and future. In M.W. Riley, R. Kahn and
    A. Foner (Eds.), *Age and structural lag* (pp. 15–36). New York: John Wiley and Sons.

Rosow, I. (1967). *Social integration of the aged*. New York: Free Press.

Rowles, G., and Watkins, J. (1993). Elderly Migration and Development in Small Communities. *Growth and Chance, 24*(4), 509–538.

Serow, W. (2003). The economic consequences of retiree concentrations: A review of North American studies. *The Gerontologist, 43*(6), 897–903.

Silverstein, M., and Angelelli, J. (1998). Older parents' expectations of moving closer to their children. *Journal of Gerontology: Social Sciences, 53*(3), S153–S163.

Skelley, B. (2004). Retiree-attraction policies: Challenges for local governance in rural regions. *Public Administration and Management, 9*(3), 212–223.

Sofranko, A., Fliegel. F., and Glasgow, N. (1983). Older urban migrants in rural settings: Problems and prospects. *International Journal of Aging and Human Development, 16*(4), 297–309.

Stallman, J., Deller, S., and Shields, M. (1999). The economic impact of aging retirees on a small rural region. *The Gerontologist, 39*(5), 599–610.

Thoits, P., and Hewitt, L. (2001). Volunteer work and well-being. *Journal of Health and Social Behavior, 42*(2), 115–131.

Walzer, N., and Deller, S. (1991). Economic change and the local public sector. In N. Walzer (Ed.), *Rural community economic development* (pp. 65–82). Westport, CT: Praeger Publishers.

Wethington, E., Moen, P., Glasgow, N., and Pillemer, K. (2000) Multiple roles, social integration and health. In K. Pillemer, P. Moen, E. Wethington and N. Glasgow (Eds.), *Social integration in the second half of life* (pp. 48–71). Baltimore: Johns Hopkins University Press.

Wiseman, R. (1980). Why older people move. *Research on Aging, 2*(1), 141–154.

Young, F. (2004). Community structure and population health: The challenge of explanation. In N. Glasgow, L. Morton and N. Johnson (Eds.), *Critical issues in rural health* (pp. 261–270). Ames, Iowa: Blackwell Publishers.

Young, F., and Glasgow, N. (1998). Voluntary social participation and health. *Research on Aging, 20*(3), 339–362.

INDEX

Page numbers with *f* indicate figures. Page numbers with *t* indicate tables. Endnote numbers follow page numbers with *n*.

201

Rural retirement migration
  choice of rate for, 24–25
  economic impact of, 39–41
  as focus of research, 9–10
  micro- and macro- aspects of, 3
  migration of younger persons vs., 46n3
  motivation for study of, 1–2
  population aging and, 38
  retiree attraction programs, 41–43
  socio-economic status and, 39
  socio-economic status of longer-term older
    residents and, 180
  to unplanned vs. planned communities,
    179–180
  See also Migrants, older; Migration;
    Retirement migration

St. Andrews Village, Maine, 180
  See also Lincoln County, Maine
Savings
  importance of, 101
  See also Economic security
Schacter, J., 24
Seasonal migration
  arts and cultural organizations and, 165
  Cornell Retirement Migration Project and,
    99–100, 113n13
  in developed countries, 11
  to Gila County, Arizona, 153
  to Leelanau County, Michigan,
    154, 161
  to Lincoln County, Maine, 156–157
  secondary tourism and, 162, 183
  to Transylvania County, North Carolina,
    157–158
  See also Migration; Rural retirement
    migration
Secondary tourism, seasonal migration
  and, 162, 183
Seidman, L. W., 40–41
Selectmen
  definition of, 175n1
  interviews with, 144, 146, 177
  in Lincoln County, Maine, 157
Self-perceived health
  factors associated with changes in, 134, 135t,
    136–137
  as variable in health status measurement, 129
  See also Health status

Senior center activities, 130
Serow, W., 5–6, 39, 41, 186
Service providers, *see* Aging services providers
Services-dependent areas
  RRDs and, 34, 34t
  younger persons' use of public
    services in, 41
  See also Public services
Severinghaus, J., 42
Sheppard, J., 11
Shields, M., 10, 39, 41, 45, 46n9,
    184, 186, 191
Silverstein, M., 106, 180
Singer, J. D., 128
Size of population
  as migrants' reason for leaving origin, 105
  migration decision making and, 94
  of retirement community, RRD designation
    and, 85
Skelley, B. D., 41, 45, 191
Sleeping Bear Dunes National Lakeshore,
    Michigan, 154
Smart growth, impact of older in-migrants
    on, 169
SMSA (Standard Metropolitan Statistical
    Areas), *see* Metropolitan areas
Snow birds, *see* Seasonal migration
Soar, J., 11
Social and civic engagement
  by older in-migrants, 163, 164, 165–166, 187
  population aging and, 2
  See also Social integration; Social networks
Social and economic attributes
  adaptation to population changes and,
    189–190
  of RRDs, 44
Social integration
  background and significance, 119
  conceptualization of, as framework for
    research, 13
  conclusion on health status and, 137–139
  data and methods for study of, 120–121
  definitions of, 118–119
  hypothesis of health status and, 128–129
  independent variables in measurement
    of, 129–130
  introduction to, 16–17, 117–118
  as key to successful aging, 185–186
  likelihood for older persons of, 121

Springer Series on Demographic Methods and Population Analysis

springer.com

Printed in the United States
117896LV00003B/259/P